# VITAL SIGNS
## 2002

# Other Norton/Worldwatch Books

*State of the World 1984* through *2002* (an annual report on progress toward a
  sustainable society)
  Lester R. Brown et al.

*Vital Signs 1992* through *2001* (an annual report on the environmental trends that are
  shaping our future)
  Lester R. Brown et al.

*Saving the Planet*
  Lester R. Brown
  Christopher Flavin
  Sandra Postel

*How Much is Enough?*
  Alan Thein Durning

*Last Oasis*
  Sandra Postel

*Full House*
  Lester R. Brown
  Hal Kane

*Power Surge*
  Christopher Flavin
  Nicholas Lenssen

*Who Will Feed China?*
  Lester R. Brown

*Tough Choices*
  Lester R. Brown

*Fighting for Survival*
  Michael Renner

*The Natural Wealth of Nations*
  David Malin Roodman

*Life Out of Bounds*
  Chris Bright

*Beyond Malthus*
  Lester R. Brown
  Gary Gardner
  Brian Halweil

*Pillar of Sand*
  Sandra Postel

*Vanishing Borders*
  Hilary French

# VITAL SIGNS
## 2002

*The Trends That Are Shaping Our Future*

## WORLDWATCH INSTITUTE

Janet N. Abramovitz

Erik Assadourian

Lester R. Brown

Jessica Dodson

Seth Dunn

Christopher Flavin

Hilary French

Gary Gardner

Brian Halweil

Kathleen Huvane

Ann Hwang

Janet Larsen

Nicholas Lenssen

Lisa Mastny

Anne Platt McGinn

Danielle Nierenberg

Sandra Postel

Michael Renner

David M. Roodman

Payal Sampat

Uta Saoshiro

Michael Scholand

Molly O. Sheehan

Linda Starke, *Editor*

 *In cooperation with the United Nations Environment Programme*
UNEP

W.W. Norton & Company
New York   London

The text of this book is composed in ITC Berkeley Oldstyle with the display set in Optima

Composition by the Worldwatch Institute; manufacturing by the Haddon Craftsmen, Inc.
Book design by Elizabeth Doherty.

ISBN 0-393-32315-3 (pbk)

W.W. Norton & Company, Inc.
500 Fifth Avenue, New York, NY 10110
W.W. Norton & Company Ltd.
75/76 Wells Street, London W1T 3QT

1234567890

## Coming Soon — a new CD-ROM

Later this year, Worldwatch will be offering the data from all the Figures in this book on a CD-ROM. This valuable research and reference tool will display its data in a spreadsheet format with color enhanced graphics. In addition, it will include user-friendly software for both PC and Macintosh computer systems that will allow you to browse, search full text, print, or export a rich collection of information. For more details or to order, please call our Customer Service center at (888) 544-2303 or (570) 320-2076. You can also find information on the new CD-ROM by going to our Website at <secure.worldwatch.org/cgi-bin/wwinst/>.

**Visit us on the Web at www.worldwatch.org**

# TABLE OF CONTENTS

## PART TWO: SPECIAL FEATURES

# ACKNOWLEDGMENTS

As we strive to make each edition of *Vital Signs* more useful than the last, we thank you, our readers. From educators and reporters to government officials and concerned citizens, many of you have taken the time to tell us how you are using this book to make a difference in the world. A lecturer at Florida State University is mining *Vital Signs* for an "Environmental Minute" TV show. On the other side of the world, the Director of the Global Environment Program at Vietnam National University has sent copies of the Vietnamese edition to senior government officials. An editor of a weekly publication in Latin America says she uses *Vital Signs* as a primary resource, "alongside my Webster's dictionary, AP style book, and *New York Times Almanac*." And to keep colleagues informed of crucial global trends, a Scottish parliamentarian asked the Parliament's library to order *Vital Signs*.

We also acknowledge a similarly rich array of outside experts who lent their time to read early drafts or provide essential data for this year's book: Tracey Axelsson, Dirk Bake, Judy Ballard, Nalini Basarajav, Nancy Birdsall, Sebastian Bizarri, Nils Borg, Mario Borsese, Barry Bredenkamp, Robert Bryant, Anil Cabraal, Judy Canny, Mark Chase, Jennifer Clapp, Colin Couchman, Charlie Craig, Adele Crispoldi, Stacy Davis, John Dearing, Lisa DiRosa, Peter du Pont, Michael Eckhart, Robert Engelman, Edward Finlay, Sharon Flesher, Satoshi Fujino, Lew Fulton, Michael Graber, Pierre Graftieaux, Maria Graschew, Roland Griffiths, Trudi Griggs, Steve Haley, Suzanne S. Hurd, Mike Jacobson, Doris Johnsen, Tim Kelly, Alison H. Kranias, Jonathan Krueger, Joe La Dou, Kimmo Laine, Benoit Lebot, David Leonhardt, Arne Lindelien, David Ludwig, Angus Maddison, Peter Markusson, Paul Maycock, Kevin McLaughlin, James McMahon, Lai Meng, Gladys C. Moreno Garcia, Bill Mott, Gerald Mutisya, Patricia Noel, Linda Novick, Peter Novy, Kate O'Neill, Brendan O'Neill, Jim Paul, Steve Plotkin, Jim Puckett, Morten Rettig, Benoît Robert, John Rodwan, Sara Scherr, Lee Schipper, J-Baptiste Schmider, Kira Schmidt, Wolfgang Schreiber, Susan Shaheen, Vladimir Slivyak, Margareta Sollenberg, Kathryn Steinberg, Diane Striar, Russell Sturm, Elizabeth Sullivan, Cornelia Thoma, Lou Thompson, Jennifer Thorne, Paul Waide, Reg Watson, Arthur Westing, Paul Steely White, Neville Williams, Stanley Wood, Wilson Wood, Karen Worminghaus, Cathy Wright, Marc Xuereb, and Dorothy Zbicz.

In addition to donations of expertise, we rely on the financial support of more than 2,000 individuals we count as Friends of Worldwatch. Special thanks to our Council of Sponsors—Adam and Rachel Albright, Tom and Cathy Crain, Roger and Vicki Sant, Robert Wallace and Raisa Scriabine, and Eckart Wintzen—and our Benefactor, Hunter Lewis. And we are grateful for the steadfast support that The W. Alton Jones Foundation has provided to *Vital Signs* since the book's inception.

This is the second edition we have produced in cooperation with the U.N. Environment Programme (UNEP). We appreciate in particular the help provided by Marion Cheatle and Tim Foresman at UNEP, who have made the relationship such a comfortable as well as a beneficial one. In addition, we benefited greatly from the careful review provided for UNEP by Mirjam Schomaker, who helped us with accuracy and clarity.

Numerous foundations support our general research program, which underlies all our publications and lets Worldwatch speak with an independent voice: Geraldine R. Dodge Foundation, The Ford Foundation, Richard & Rhoda Goldman Fund, The William and Flora Hewlett Foundation, The Frances Lear Foundation, Steve Leuthold Family Foundation, The John D. and Catherine T. MacArthur Foundation, Charles Stewart Mott Foundation, The Curtis and Edith Munson Foundation, The David and Lucile Packard Foundation, The Shenandoah Foundation, The Summit Foundation, Surdna Foundation, Inc., Turner Foundation, Inc., the Wallace Global Fund, the Weeden Foundation, and The Winslow Foundation.

We are privileged to have a particularly close relationship with our U.S. publisher, W.W. Norton & Company. Amy Cherry, Lucinda Bartley, and Andrew Marasia at Norton speed the book through the publishing process. We also value our partners who publish *Vital Signs* outside the United States in 22 languages.

The in-house people and talents that Worldwatch draws on to produce *Vital Signs* are as varied as the indicators in this book. This year, the Project Team consisted of Michael Renner (Project Director), Brian Halweil, and Molly O'Meara Sheehan. As we prepared this edition, crucial support was provided by Institute stalwarts Barbara Fallin and Suzanne Clift; our business and development team of Adrianne Greenlees, Elizabeth Nolan, Kevin Parker, Mary Redfern, and Cyndi Cramer; and our communications team of Dick Bell, Leanne Mitchell, Patrick Settle, Sharon Lapier, Niki Clark, and Susanne Martikke. For the hardest-to-find reports and data sets, authors rely on our research librarian Lori Brown, assisted by Jonathan Guzman, and on Joseph Gravely in our mailroom.

This year, our regular research staff was bolstered by a network of Worldwatch alumni: Ann Hwang, Janet Larsen, Nick Lenssen, and Mike Scholand. While not full-time staffers, Worldwatch Board Member Lester Brown and Senior Fellow Sandra Postel, assisted by Katie Blake, also made key contributions. An especially talented crew of interns, including Jessica Dodson, Kathleen Huvane, and Uta Saoshiro, found time to draft pieces of their own while assisting senior researchers, as did Erik Assadourian, who has since come on board as a full-time researcher. Arriving late in December, our newest intern, Meghan Crimmins, pitched in during crunch time.

Finally, we thank two individuals at the core of this book. Independent editor Linda Starke held authors' feet to the fire, turning dozens of drafts submitted by 23 nearby and far-flung authors into polished prose at breakneck speed. Working under the most intense deadline pressure, Art Director Eizabeth Doherty maintained her creative spark to make *Vital Signs* both better-looking and easier to understand. Several of the photos Liz selected for this edition are from Photoshare, the online photo database of the Media/Materials Clearinghouse at the Johns Hopkins University Population Information Program at <www.jhuccp.org/mmc>. We are sad to note that this is Liz's final *Vital Signs*. Since September 1996 Liz has brought considerable talents and an untiring spirit to six editions of this book. We wish her well in her new endeavors.

Information on our CD-ROM, which contains the data used to prepare all of the Figures in this book, can be found on page 6. Let us know if you have ideas of other trends we can cover. Please contact us by e-mail (worldwatch@worldwatch.org), fax (202-296-7365), or regular mail.

*Vital Signs* Project Team
March 2002

Worldwatch Institute
1776 Massachusetts Ave., N.W.
Washington DC 20036

# PREFACE

By most standards, including many of the "vital signs" catalogued in this book, the past year would be classified as an *annus horribilis*. A year that began with economic recession and heavily publicized food safety scares was later marked by violent outbreaks of ethnic conflict and the most deadly single episode of terrorism the world has ever seen.

Hopes that the world had entered a period of peace and prosperity at the dawn of the twenty-first century had to be put aside as the year proceeded, amid growing awareness of the instabilities inherent in a period of accelerating change—and the web of interconnections that make people everywhere vulnerable to crises that break out anywhere.

*Vital Signs 2002* focuses not on the spectacular events that dominated news coverage of the past year but on the deeper, more chronic trends that define the health of people and the planet—and that provide the context for the crises that command public attention. These trends now point to a dangerous instability, one that can only be righted by concerted efforts to create a more secure and sustainable world.

The fact that 1.2 billion people live on less than $1 a day—a figure roughly unchanged even after a decade of phenomenal economic growth in much of the developing world—is clearly undermining stability in some societies. And rapid economic growth has created a rising gap between rich and poor in many countries, another force of instability.

So long as 3 million people die yearly from AIDS, 100–150 million suffer from asthma, and 2.4 billion lack basic sanitation—all documented in the pages that follow—it is hard to imagine that we can achieve a stable or secure world.

Growing instability is seen in the natural world as well. The year 2001 was the second warmest on record, joining a list of the 10 warmest years in the last century—all of which have occurred since 1990. Carbon dioxide, the leading greenhouse gas, continues to build up in the atmosphere as carbon emissions reached a new high.

On the ground, an estimated 150–300 million hectares of cropland—10–20 percent of the world total—is now degraded. More than 2 billion people live in water-stressed countries in which water supplies are insufficient to meet food, industrial, and household needs.

When world leaders gather at the World Summit on Sustainable Development in Johannesburg, South Africa, they will face no shortage of challenges. Indeed, the need for a global action plan on the interlinked problems of environmental decline and human poverty has never been as evident as it is this year.

While the problems facing the world in Johannesburg are daunting, *Vital Signs 2002* also offers encouraging evidence that national policy and even human behavior can change in response to new threats—and that sometimes solutions emerge that no one would have expected.

Who would have guessed a decade ago, for

example, that the world leader in producing the efficient compact fluorescent light bulbs pioneered in Europe and the United States would be China? Or that wind power would become the world's fastest-growing energy source—with annual additions to generating capacity on the verge of overtaking hydropower? And who would have imagined that the fastest-growing transportation trend in industrial countries would be car *sharing*, an alternative to private ownership that reduces the temptation to overuse the automobile?

As these few examples suggest, change can sometimes happen quickly, and it is most effective when it involves both the innovative capacities of private citizens and companies and the societal goals and incentives that are the province of governments and international agencies. The Johannesburg Summit offers an opportunity to move forward with implementation of agreements now in place, pursuing strategies that will provide economic opportunities at the same time that they solve environmental problems.

The Worldwatch Institute and the United Nations Environment Programme are both convinced that change is possible—and that an informed public is the first ingredient of productive change. We hope that *Vital Signs 2002* will provide some of the information that people and their leaders need to make wise decisions.

Christopher Flavin
President
Worldwatch Institute

Klaus Töpfer
Executive Director
United Nations Environment Programme

# VITAL SIGNS
## 2002

## TECHNICAL NOTE

Units of measure throughout this book are metric unless common usage dictates otherwise. Historical population data used in per capita calculations are from the Center for International Research at the U.S. Bureau of the Census. Historical data series in *Vital Signs* are updated each year, incorporating any revisions by originating organizations.

Data expressed in U.S. dollars have for the most part been deflated to 2000 terms. In some cases, the original data source provided the numbers in deflated terms or supplied an appropriate deflator, as with gross world product data. Where this did not happen, the U.S. implicit gross national product (GNP) deflator from the U.S. Department of Commerce was used to represent price trends in real terms.

# OVERVIEW

## Making the Connections

*Michael Renner*

In the aftermath of 11 September 2001, many people have said that the terror attacks changed the world in fundamental ways. It may be more appropriate to say that the shocking events of that day were a dramatic wake-up call—a catalyst for undertaking a critical reassessment of the state of affairs on our globe, and of the underlying conditions that feed desperation, fuel resentment, and breed violence. A candid appraisal reveals widening disparities between rich and poor, mounting health challenges, battered ecosystems, and persistent social and political conflicts. Yet there are also many opportunities for positive change through the promotion of social justice and environmental health, international cooperation, technological innovation, and greater prudence in the pursuit of human ingenuity. Many of those topics will be addressed in Johannesburg in August–September at the World Summit on Sustainable Development—an ideal time to capitalize on the opportunities for change.

*Vital Signs 2002* offers information on a broad range of issues critical to putting the world on a more just, ecologically resilient, and ultimately peaceful trajectory. It brings together a careful selection of topics, seen through the lens of global equity and sustainability. As in previous editions, *Vital Signs* covers a range of basic and long-established indicators such as gross economic product and trade flows, population growth, grain production, fossil fuel consumption, automobile manufacturing, and roundwood production. And it continues to document alternative indicators of ever-growing significance, like wind and solar power development, bicycle production, carbon emissions, chlorofluorocarbon (CFC) use, and the growth of biotechnology.

But in recognition of the many issues critical to sustainability, new topics are also covered in *Vital Signs 2002*. Roughly one third of the book addresses issues not covered earlier, including sugar crops, soft drink consumption, oil spills, hazardous waste trade, ecolabeling, appliance efficiency standards, car-sharing, urban sprawl, asthma, mental health, the cruise industry, transboundary parks, teacher shortages, and gender-based violence.

Among the most promising developments documented in *Vital Signs 2002* are the surging sales of efficient compact fluorescent lamps (CFLs, with an estimated 1.8 billion in use worldwide), the continued rapid expansion of wind and solar-generated electricity, the steady decline in the amount of oil spilled accidentally, and the ongoing reduction in production of ozone-destroying chemicals. Other encouraging developments are the decreasing metals intensity of the world economy, the growing reliance on transboundary parks as tools for biodiversity conservation and peace- and confidence-building, the expansion of commercial forest areas that have been certified as well-managed, reductions in the number of active armed conflicts, and progress in curtailing reliance on landmines.

On the downside, there is ongoing forest loss in the tropics, the threat of extinction for many freshwater species, the relentless generation of huge amounts of hazardous waste, the continued expansion of the car-centered transportation system, the massive spread of HIV infections, runaway consumption of sugar and soft drinks, widespread teacher shortages, an epidemic of violence against women, and declining foreign aid.

The impacts of some of the trends documented in *Vital Signs* are self-evident. Others may be less clear-cut. For instance, there is nothing intrinsically wrong with increased cocoa production, but reports of children being forced to work in slavery-like conditions in some areas add a negative tint to this trend. Most economists regard growing car production as a positive development because of job creation and enhanced mobility. But the rising costs of a car-centered transportation system—from air pollution and carbon emissions to urban sprawl and the fatalities and injuries from traffic accidents—suggest a more negative assessment.

Qualitative assessments of Earth's vital signs are of necessity subjective in nature, the result of different sets of values, philosophies, expectations, and goals. The proverbial glass can be seen as half full or half empty. Readers may draw their own conclusions.

## CONNECTIONS

Although each individual item in this book was written as a stand-alone piece, the intention is to encourage readers to engage in cross-cutting comparisons among related issues. The contents of this year's *Vital Signs* can be grouped in a variety of topic clusters. This overview looks at three such clusters—energy, climate, and transportation; land, water, and food; and the impact of technology. These are only some of many cross-cutting issues to emerge. Readers might want to do their own comparisons of material in this book and draw linkages and conclusions that are germane to their work and interests.

Due to expanding trade, travel, and communications networks, the world has become ever more interlinked, so that events in far-flung places affect millions elsewhere on the planet. This is as true for economic and political issues as for social and environmental ones.

Other connections are equally crucial and yet too often remain unacknowledged. When millions of motorists turn on their cars in the morning on their way to work, they may not be aware that the simple act of driving is contributing to the unraveling of the climate system, thus helping to cause or worsen floods in Bangladesh, mudslides in Central America, or droughts in parts of Africa. At the furniture store, consumers may buy products made from wood harvested in destructive logging operations that threaten the livelihoods of indigenous populations. As these two simple examples illustrate, no society lives in isolation in this interlinked world. Oceans and other natural barriers are no longer insurmountable; borders are far from impermeable. The challenge in a world of nation-states of different size and power is to devise ways to maximize the benefits and minimize the damage from the globalization now being experienced.

## ENERGY, CLIMATE, AND TRANSPORTATION

An understanding of the manifold and complex connections that characterize the modern world is increasingly critical. Energy plays a particularly important role. The global economy has long depended on the availability of abundant supplies of cheap energy, particularly from the politically volatile Persian Gulf region. Maintaining access to oil at all cost has been a central tenet of economic and military policies of western industrial countries. But this policy has contributed to repeated upheavals in the Middle East. The energy status quo not only implies continued instability for the world economy and for world peace, it also has grim consequences for the stability of the global atmosphere. (See Figure 1.)

Fossil fuel consumption and carbon emissions each rose more than 1 percent in 2001, reaching new peaks. (See pages 38–39 and 52–53.) Global temperatures have been on the

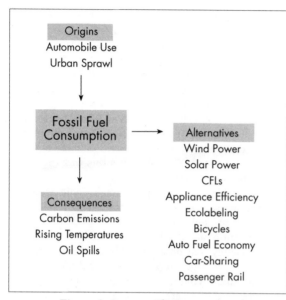

**Figure 1: Energy, Climate, and Transportation Connections**

upswing during the past half-century, and land and ocean measurements show that 2001 was the second-warmest year on record since the late nineteenth century. Not surprisingly, 2001 brought several episodes of abnormal weather, including an above-average number of hurricanes and tropical storms in the north Atlantic basin; severe flooding in Viet Nam, Siberia, and different parts of Africa; and devastating droughts in Iran, Afghanistan, Pakistan, the Horn of Africa, Brazil, northern China, North Korea, and Japan. (See pages 50–51.)

To quench the industrial world's thirst for fossil fuels, tankers transport some 107 million tons of oil each day. Oil tankers are a leading source of oil spills, though pipelines, production wells, storage facilities, and refineries are important sources as well. The good news is that a variety of safety measures have helped reduce oil spills from civilian operations. The amount of oil lost in 2000, almost 50,000 tons, was the lowest since continuous recordkeeping began in 1968. Still, even small amounts of oil can do major damage if an accident occurs in or near a fragile ecosystem. (See pages 68–69.)

Car-centered transportation is playing a major role in the world's voracious appetite for fossil fuels. This is particularly the case in sprawling urban areas where long travel distances render biking and public transport almost impossible while making reliance on cars a daily inevitability. During the 1990s, road transportation was the fastest-growing source of carbon emissions from fuel burning. There are now 555 million passenger vehicles on the world's roads, and factories churn out about 40 million new cars each year. (See pages 74–75.) Although car fuel economy is again improving after having stagnated for many years, it remains far short of technical possibilities. And in the United States, which has slightly more than a quarter of the world's cars, there is little prospect of significant improvement over the next decade. (See pages 152–53.)

Passenger-kilometers traveled by rail have stagnated since the late 1980s, and rail continues to lose out to travel by car and airplane. (See pages 78–79.) Meanwhile, global production of bicycles has recovered from a slump, topping 100 million units in 2000 for the first time since 1995. But the bicycle industry continues to struggle. (See pages 76–77.) Particularly in Europe, an alternative approach is rapidly gaining adherents. Car-sharing is attracting rising numbers of people who do not see a need to own a car themselves. Such ventures offer social and environmental benefits to cities. (See pages 150–51.)

Headway is being made in some other ways to reduce energy use. Compact fluorescent lamps are longer-lasting and far more energy-thrifty than conventional incandescent light bulbs. Sales of CFLs worldwide grew 15 percent in 2001 alone, and have increased more than 13-fold since 1988. (See pages 46–47.) Efficiency standards for domestic appliances have been initiated in 43 countries worldwide, and have helped eliminate more energy-thirsty models from the market. (See pages 132–33.) Consumers can make more responsible purchasing decisions by relying on ecolabeling that guides them toward more-efficient and environmentally benign goods and services. (See pages 124–25.)

Making more efficient use of fossil fuels is only part of the equation. An equally important task is to promote alternative sources of energy. Wind and solar power have been growing rapidly in recent years, and use of each expanded by more than 30 percent in 2001 alone. (See pages 42–45.)

## LAND, WATER, AND FOOD

A number of critical connections also exist in the realm of food and agriculture. Arable land and water for agriculture are among the most critical resources for human well-being and survival, no matter the technological prowess of a society. Yet freshwater resources are often tapped beyond sustainable rates and many cropland areas are pushed to the limits. Although the global grain harvest is near peak levels, farmers and consumers confront a number of serious quantitative and qualitative challenges. (See Figure 2 and pages 26–27.)

An estimated 10–20 percent of the world's 1.5 billion hectares of cropland are degraded to some degree, the result of excessive tillage and fertilizer use, inappropriate land use, removal of vegetation, and overgrazing. In the developing world, the pace of decline has accelerated during the past 50 years to the point where a quarter of the farmland suffers from degradation. Worldwide, farmland degradation has reduced cumulative food production by an estimated 13 percent over the last half-century. (See pages 102–03.)

Urban expansion eats into prime agricultural land, particularly in the case of cities that are characterized by a pattern of sprawl. For instance, although only 3 percent of the U.S. land surface is urbanized, the most productive soils are often developed first as cities

expand. In fact, more than 1 million hectares of arable land in the United States are paved over each year. In China, the figure is 200,000 hectares. (See pages 152–53.)

Another common factor in farmland degradation is salinization—a buildup of salt that occurs when excess irrigation water evaporates. Salinization can hurt yields and even force the abandonment of irrigated land. Today, about 20 percent of the world's 274 million hectares of irrigated land are damaged in this way. (See pages 34–35 and 102–03.)

Improved irrigation efficiency could avoid these problems and raise farm yields, but at the moment, inefficient methods are used on 90 percent of artificially watered fields. Greater efficiency is also important because growing water shortages in Africa, Asia, and the Middle East are forcing an increasing number of countries to rely on grain imports. By 2015, with rising water shortages and populations, a projected 40 percent of humanity will live in water-stressed countries, putting increasing pressure on global grain supplies. Making low-cost, efficient irrigation available to poor farmers will be key to alleviating hunger and malnutrition. (See pages 34–35 and 148–49.)

More efficient water use is also essential to save many freshwater species from extinction and to preserve the valuable ecological services

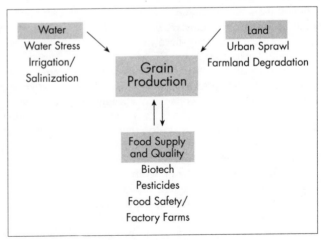

**Figure 2: Land, Water, and Food Connections**

they provide, such as filtering and cleansing water supplies and mitigating floods and droughts. The habitat of these species is increasingly under assault by dams, river diversions, pollution, and the introduction of nonnative species. Almost 80 percent of the largest river systems in North America, Europe, and the former Soviet Union are moderately or strongly altered by dams, reservoirs, diversions, and irrigation systems, and similar challenges are now arising in the developing world. (See pages 106–07.)

Farming and other components of food production have become industrialized, resource-intensive systems. On the input side, pesticide use (two thirds of it in agriculture) has grown 15-fold since 1950 but imposes a terrible toll, poisoning 3 million people severely and killing 220,000 each year. Meanwhile, farmers confront increasing pesticide resistance. (See pages 126–27.)

For consumers, food quality ranks among the most widespread health concerns. Foodborne diseases strike 30 percent of the population in industrial countries each year, but people living in developing countries bear a more frightful burden due to a wide range of hazards and inadequate prevention and treatment. Though lack of household hygiene is a factor, many problems begin far earlier. Livestock in many modern factory farms, for instance, are often raised in crowded, unsanitary conditions, which promotes food-borne illnesses. (See pages 138–39.)

## THE HAZARDS OF HIGH-TECH AND OLD TECH

Humanity is confronting some of the broad boomerang effects of modern technology. The unintended consequences of what once seemed technological marvels can entail severe threats to human health and well-being. Nuclear power, at first considered too cheap to meter, is bequeathing the unwanted long-term "gift" of radioactive waste. (See pages 40–41.) Chlorofluorocarbons, for decades judged ideal for refrigerating, air-conditioning, and a host of

other purposes, turned out to be efficient killers of the atmospheric ozone layer that protects life on Earth from deadly ultraviolet radiation. Though CFC production is now down sharply, it may take a half-century for the ozone layer to heal completely. (See pages 54–55.)

Modern industrial life is characterized by the generation of substantial amounts of hazardous waste—both in traditional industries such as metals mining and processing, petrochemicals, pesticides, and plastics manufacturing and in newer, more high-tech sectors. Some 300–500 million tons of heavy metals, solvents, toxic sludge, and other wastes accumulate each year. (See Figure 3 and pages 66–67 and 112–13.)

The semiconductor industry has undergone explosive growth in the past two decades. In 2001, some 60 million transistors—the tiny components used to build semiconductor chips—were manufactured for each person in the world. But because of the rapid pace at which electronic products become obsolete and are being replaced, production is expected to skyrocket in coming years, to perhaps as many as 1 billion transistors per person in 2010. Yet the industry requires copious amounts of chemicals and leaves behind huge quantities of dangerous wastes. Production of a single six-inch silicon wafer results in 14 kilograms of solid waste and 11,000 liters of waste water. Workers in the industry are on the frontline of exposure and at risk of developing cancer or seeing birth defects in their children. (See pages 110–11.)

Cell phones are among the products that incorporate semiconductors. While they allow an ever more connected world and give millions of people access to phone service for the first time, discarded cell phones contribute to the growing mountain of electronics waste. And there is an ongoing, unresolved discussion surrounding possible harm to human health from the radio waves they emit. (See pages 84–85.)

More than 80 percent of the world's hazardous waste is produced in the United States

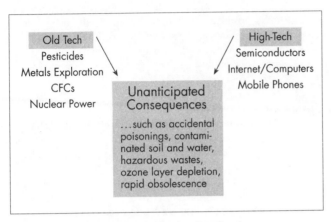

**Figure 3: Impacts of Technology**

and other industrial countries. The international community has struggled to devise and enforce rules to reduce cross-border movements in the hope of preventing poor countries from being turned into dumping grounds for the wastes of the rich. Today, about 10 percent of all hazardous waste is shipped across an international border. (See pages 112–13.)

Though much hazardous waste trade takes place among industrial countries, there are some important exceptions. The Basel Action Network, with support from other citizens' groups, found that huge quantities of computer monitors, cell phones, circuit boards, and other items from the United States end up in China, India, and Pakistan. There, they are either being dumped or the materials they contain— lead, mercury, cadmium, copper, gold, and many others—are salvaged in such crude ways as to pose a severe occupational and environmental threat. (See pages 82–83.) Separated by thousands of kilometers, beneficiaries and victims of the high-tech revolution never meet face-to-face, but the connections between them are real.

Time and again, technological innovation has kicked loose a range of unintended consequences. Depending on the situation and the

time, an overly narrow focus of scientific inquiry, excessive technological optimism, unbridled reign of the profit motive, or plain lack of foresight may lead societies to pursue technological promise with abandon, only to discover surprising side effects, unknown long-term consequences, and unanticipated feedback loops. The world is still learning to cope with the repercussions of the chemical revolution, even as it hurtles with great speed through the electronics age and plunges headlong into the biotech era.

Increasingly, the challenge for scientists, corporations, governments, and individuals is to use human inventions more judiciously— with an eye to the likely implications for equity and sustainability. That requires greater wisdom in deciding what technologies to pursue, how to mold them, and when to look for alternatives. Simply striving for the technically feasible is no longer a responsible option. Indeed, the precautionary principle—in the face of scientific uncertainty, exercise caution—becomes ever more important as our lives are increasingly permeated by the creations of human ingenuity and hubris. This is possible only with a more holistic view of the world, and a better understanding of the kinds of connections that this book explores.

# PART ONE
## Key Indicators

# Food and Agricultural Trends

Aquaculture Production Intensifies

Grain Harvest Lagging Behind Demand

Meat Production Hits Another High

Cocoa Production Jumps

Sugar and Sweetener Use Grows

Irrigated Area Rises

# Aquaculture Production Intensifies

*Anne Platt McGinn*

Global aquaculture production has grown nearly 400 percent in the past 15 years, from 7 million tons in 1984 (the first year with global data) to 33.3 million tons in 1999.[1] (See Figure 1.) Preliminary data indicate production climbed to 36.1 million tons in 2000.[2]

Aquaculture is the fastest-growing segment of food production in the world.[3] As global marine catches stagnate and even decline in some areas, aquaculture is quickly filling the gap. It now provides 31 percent of the world's food fish, up from 19 percent in 1990.[4] Globally, the value of farmed fish doubled from $24.5 billion in 1990 to $47.9 billion in 1999.[5] (By comparison, fish catches were valued at $83 billion in 1998.)[6]

Almost 9 out of 10 farmed fish in the world—some 86 percent—are now raised in Asia.[7] Farmers in China boosted output by 252 percent during the 1990s, and now contribute 68 percent of the world's farmed fish by volume and nearly half of its value.[8] (Unofficial reports indicate, however, that China has inflated its production data.)[9] India is a distant second in terms of output, followed by Japan, Indonesia, and Bangladesh.[10] By value, Japan, India, Indonesia, and Thailand round out the top five producers in the world.[11]

Links:
pp. 106, 138

Chile posted the largest percentage gain in the last decade, with production jumping more than 700 percent—from 32,447 tons of fish in 1990 to 274,216 tons in 1999.[12] Farmed salmon and trout account for nearly 85 percent of Chile's output.[13]

Some 220 fish species are now cultivated in captivity, although 20 species account for 90 percent of world production.[14] From 1990 to 1999, world production of farmed carp, tilapia, and other freshwater fish nearly tripled, and now accounts for 56 percent of total output.[15] (See Figure 2.) These low-value species are generally raised and consumed locally.

In contrast, high-end species such as shrimp and salmon are grown primarily for export to Japan, North America, and Europe. Production of farmed shrimp and salmon roughly doubled during the 1990s, to just 8 percent of the total, but these two species now account for 24 percent of the value of world aquaculture.[16]

The net trade earnings from captured and cultured fish in developing countries grew from $5.2 billion in 1985 to $15 billion in 1998.[17] Developing countries now earn more foreign exchange from exported fish products than from coffee, tea, rice, and rubber exports combined.[18]

Rapid growth in aquaculture has raised a number of concerns, however. Disease outbreaks have taken a stiff toll, especially where high numbers of a single species are raised in small areas. In 1999, Ecuador lost nearly $500 million in export earnings due to a catastrophic outbreak of white spot virus in farmed shrimp.[19]

Another concern is aquaculture's growing appetite for wild fish. Carnivorous fish such as salmon and shrimp are typically fed high-protein pellets made from a combination of fishmeal and plant-based proteins. (Small pelagic species, such as anchovy, herring, and menhaden, are used to produce fishmeal.) Today, increasing numbers of farmers are replacing an entirely plant-based diet for omnivorous and herbivorous fish with feed pellets, to induce faster growth and weight gain.[20] As a result, the share of world fishmeal dedicated to aquaculture has increased from 10 percent in 1988 to 35 percent in 1998.[21] During that time, global fishmeal output remained steady while the share for poultry and cattle declined.[22]

In contrast, marine-raised mollusks need few artificial inputs because they feed on nutrients from the surrounding water. In 1999, cultured oysters and clams commanded 14 percent of the value of global aquaculture.[23] Some experts are encouraging displaced fishers to adopt environmentally sound aquaculture to help generate income. For example, farmers can cultivate species that fetch high prices on international markets, such as oysters for pearls and giant clams for the aquarium industry.[24] But export-driven aquaculture does not eliminate the importance of raising fish for local consumption, a growing need in many food-deficit countries.

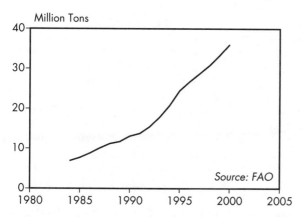

Figure 1: World Aquaculture Production, 1984–2000

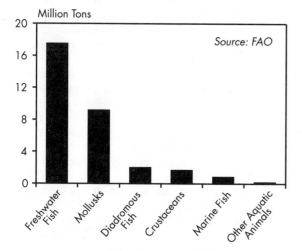

Figure 2: World Aquaculture Production by Major Species Groups, 1998

## World Aquaculture Production, 1984–2000

| Year | Production (million tons) |
|------|------------|
| 1984 | 6.9 |
| 1985 | 7.7 |
| 1986 | 8.8 |
| 1987 | 10.1 |
| 1988 | 11.2 |
| 1989 | 11.7 |
| 1990 | 13.1 |
| 1991 | 13.7 |
| 1992 | 15.4 |
| 1993 | 17.8 |
| 1994 | 20.8 |
| 1995 | 24.5 |
| 1996 | 26.8 |
| 1997 | 28.7 |
| 1998 | 30.8 |
| 1999 | 33.3 |
| 2000 (prel) | 36.1 |

Source: FAO, Aquaculture Production Statistics 1984–93 and Fishery Statistics: Aquaculture Production.

# Grain Harvest Lagging Behind Demand

*Lester R. Brown*

This year's world grain harvest, estimated at 1,843 million tons, is up slightly from last year's poor harvest of 1,836 million tons.[1] (See Figure 1.) It is, nonetheless, a depressed harvest—40 million tons below 1997's record 1,880 million tons.[2]

Grain production per person worldwide this year totals 299 kilograms, down from the peak of 342 kilograms in 1984.[3] (See Figure 2.) This 14-percent decline since 1984 contrasts with a 38-percent gain from 1950 to 1984, a period of widespread progress in reducing hunger and malnutrition worldwide.[4]

The poor harvests of the last two years are a result of weak world prices for grain, of

Links: pp. 34, 102, 126, 134

drought stretching from the Middle East through central Asia and across northern China, and of spreading shortages of irrigation water. Prices will recover and the drought will end, but irrigation water shortages will worsen as population growth outruns the water supply in more and more countries.

The longer-term worldwide drop in grain production per person has been concentrated in Africa, Eastern Europe, and the former Soviet Union.[5] In Africa, soil degradation and aridity have constrained gains in food production. Limited gains or declines in grain output, coupled with the fastest population growth of any continent, have increased hunger and malnutrition.[6] Economic decline in the former Soviet Union and Eastern Europe following economic reforms and the breakup of that large nation a decade ago greatly reduced both grain production and consumption.[7]

China, the world's largest grain producer, is primarily responsible for the decline in grain-harvested area in the last two years that has lowered the world grain harvest so dramatically.[8] While world output was dropping 30 million tons in the last two years, China's grain harvest shrunk by 53 million tons, more than offsetting modest gains elsewhere.[9]

Among the forces shrinking China's grain harvest are severe drought in the north during the last two years, spreading irrigation water shortages as aquifers are depleted and as water is diverted to cities, and a lowering of support prices.[10] In a country dependent on irrigated land for 70 percent or more of its grain, water shortages are fast becoming a security issue.[11]

In 1994, in an ambitious and initially successful effort to be self-sufficient, China raised grain support prices by 40 percent.[12] Unfortunately, the drain on the treasury was too great, so the support prices were lowered in 2000 and 2001, dropping close to world market levels.[13] As grain prices have fallen over the last three years, the area planted to grain has shrunk by 10 percent.[14]

China has absorbed the harvest shortfall by drawing down stocks, but there are signs that supplies are now tightening.[15] If this huge nation, with a population equal to that of India and the United States combined, has another large harvest shortfall, it will likely have to import substantial quantities of grain to maintain food price stability.

Among the three major grains, the harvest of the two food grains—wheat and rice—each dropped in 2001 from the previous year.[16] (See Figure 3). Corn, used mostly as a feed grain for livestock, poultry, and fish, edged out wheat again as the world's leading grain.[17]

Although world grain production was down during the last two years, consumption continued to rise.[18] Grain use exceeded production by 35 million tons in 2000 and by 51 million tons in 2001.[19] The excess of production over consumption dropped grain stocks as a share of consumption to 23 percent—one of the lowest levels in two decades.[20]

If world grain demand continues to grow during 2002 at the same pace as the last decade—16 million tons a year—then this year's harvest will have to jump by 70 million tons to avoid a further drawdown in stocks.[21]

With grain stocks at such a low level, grain market analysts will be watching the 2002 harvest closely. If it falls well short of consumption, grain prices will likely climb. Spreading shortages of irrigation water as aquifers are depleted and as water is diverted to cities are making it much harder for the world's farmers to keep up with the growth in demand.

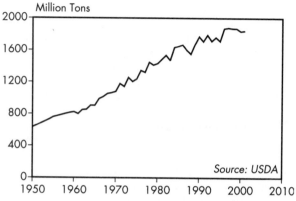

Figure 1: World Grain Production, 1950–2001

Figure 2: World Grain Production Per Person,
1950–2001

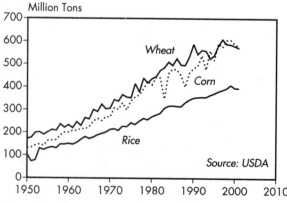

Figure 3: Wheat, Corn, and Rice Production, 1950–2001

## World Grain Production, 1950–2001

| Year | Total (mill. tons) | Per Person (kilograms) |
|---|---|---|
| 1950 | 631 | 247 |
| 1955 | 759 | 273 |
| 1960 | 824 | 271 |
| 1965 | 905 | 270 |
| 1970 | 1,079 | 291 |
| 1971 | 1,177 | 311 |
| 1972 | 1,141 | 295 |
| 1973 | 1,253 | 318 |
| 1974 | 1,204 | 300 |
| 1975 | 1,237 | 303 |
| 1976 | 1,342 | 323 |
| 1977 | 1,319 | 312 |
| 1978 | 1,445 | 336 |
| 1979 | 1,411 | 322 |
| 1980 | 1,430 | 321 |
| 1981 | 1,482 | 327 |
| 1982 | 1,533 | 332 |
| 1983 | 1,469 | 313 |
| 1984 | 1,632 | 342 |
| 1985 | 1,647 | 339 |
| 1986 | 1,665 | 337 |
| 1987 | 1,598 | 318 |
| 1988 | 1,549 | 303 |
| 1989 | 1,671 | 322 |
| 1990 | 1,769 | 335 |
| 1991 | 1,708 | 318 |
| 1992 | 1,790 | 328 |
| 1993 | 1,713 | 310 |
| 1994 | 1,760 | 314 |
| 1995 | 1,713 | 301 |
| 1996 | 1,871 | 324 |
| 1997 | 1,880 | 322 |
| 1998 | 1,872 | 316 |
| 1999 | 1,871 | 312 |
| 2000 | 1,836 | 302 |
| 2001 (prel) | 1,843 | 299 |

Source: USDA, Production, Supply, and Distribution, electronic database, December 2001.

# Meat Production Hits Another High

*Lester R. Brown*

World meat production climbed to a new high in 2001, marking the forty-first consecutive annual gain.[1] (See Figure 1.) At 237 million tons, this is up more than 2 percent over the 232 million tons of 2000.[2]

Meat production has increased more than fivefold since 1950.[3] Over this half-century, consumption per person has more than doubled, climbing from 17 kilograms to 39 kilograms.[4] (See Figure 2.)

Beef, pork, and poultry account for over 90 percent of world meat production.[5] (See Figure 3.) Most of the growth in meat output in 2001 was in pork and poultry; beef production rose less than 1 percent.[6] In fact, beef production per person has fallen by 17 percent since the historical peak in 1976.[7]

Links: pp. 26, 138

The key beef-consuming countries are the United States (12 million tons), Brazil (just over 6 million tons), and China (just under 6 million tons).[8] These three account for half of world beef consumption.[9] The European Union (EU) also weighs in with just over 6 million tons.[10]

World pork production, which overtook beef production in 1979, continued to widen the lead in 2001 as production climbed to 93 million tons, a gain of more than 3 percent.[11] Pork consumption is totally dominated by China, at 42 million tons, compared with 8 million tons in the United States, the second-ranking consumer.[12] No country dominates the consumption of a meat the way China does pork, accounting for half of world consumption.[13] The EU countries collectively eat 16 million tons of pork a year.[14]

World poultry production climbed from 67 million tons to almost 69 million tons, also gaining nearly 3 percent.[15] The steadily growing world production of poultry eclipsed that of beef in 1995, moving it into second place behind pork.[16] As of 2001, poultry consumption worldwide reached 10 kilograms per person.[17]

The United States still leads in consumption of poultry, with nearly 14 million tons, but China is closing fast at just under 13 million tons and could eclipse the United States within a few years.[18] Brazil, at just over 5 million tons

of poultry, is in third place.[19] Poultry consumption in the EU is nearly 8 million tons.[20]

Despite the uninterrupted growth in world meat consumption for more than half a century, there have been some local disruptions in recent years. For example, meat consumption in Russia declined precipitously over the last decade following economic reforms, but is now beginning to recover.[21] Meat production in the EU was disrupted a few years ago with evidence of mad cow disease and more recently by an outbreak of foot-and-mouth disease.[22] Europe is also now showing signs of recovery.[23] The identification of two cows with mad cow disease in Japan in the fall of 2001 has lowered beef consumption there.[24]

The share of world meat output that is being traded is rising, totaling nearly 16 million tons in 2001.[25] Growth in international meat trade reflects both the rising appetite for meat in middle-income countries and advances in storage and transport. Although meat is much more difficult to ship internationally than grain, the share of world meat consumption that is traded is now 8 percent, compared with 12 percent for grain.[26]

Although meat consumption is at the near-saturation point in most industrial countries, it is still growing rapidly in low- and middle-income countries, where most of the world lives. The growth in consumption in middle-income countries is evident in the most recent data. China, for example, has now emerged as the world's leading meat producer and consumer, eating some 61 million tons of meat in 2001.[27] The United States is second, at 34 million tons, and Brazil is third, at 13 million tons.[28]

While future growth in meat consumption in both the United States and Europe is expected to be limited, there is a broad potential for greater consumption not only in China and Brazil, but in other developing countries as well, such as Mexico, Thailand, and Indonesia.[29] Barring a depression in the global economy or a major disruption from livestock disease, world meat consumption is likely to continue its uninterrupted growth for the foreseeable future.

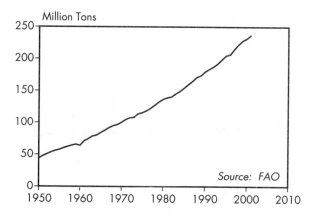

Figure 1: World Meat Production, 1950–2001

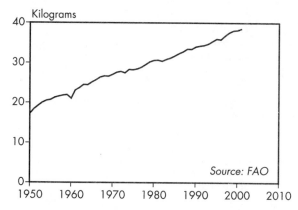

Figure 2: World Meat Production Per Person, 1950–2001

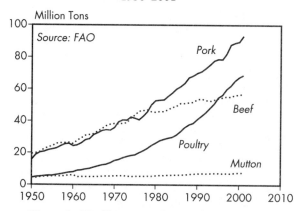

Figure 3: World Meat Production by Source, 1950–2001

## World Meat Production, 1950–2001

| Year | Total (mill. tons) | Per Person (kilograms) |
|---|---|---|
| 1950 | 44 | 17.2 |
| 1955 | 58 | 20.7 |
| 1960 | 64 | 21.0 |
| 1965 | 84 | 25.2 |
| 1970 | 100 | 27.1 |
| 1971 | 105 | 27.6 |
| 1972 | 108 | 27.8 |
| 1973 | 108 | 27.5 |
| 1974 | 114 | 28.3 |
| 1975 | 116 | 28.3 |
| 1976 | 118 | 28.5 |
| 1977 | 122 | 28.9 |
| 1978 | 127 | 29.6 |
| 1979 | 133 | 30.2 |
| 1980 | 137 | 30.6 |
| 1981 | 139 | 30.7 |
| 1982 | 140 | 30.4 |
| 1983 | 145 | 30.9 |
| 1984 | 149 | 31.2 |
| 1985 | 154 | 31.8 |
| 1986 | 160 | 32.3 |
| 1987 | 165 | 32.8 |
| 1988 | 171 | 33.5 |
| 1989 | 174 | 33.4 |
| 1990 | 180 | 34.0 |
| 1991 | 184 | 34.3 |
| 1992 | 187 | 34.4 |
| 1993 | 192 | 34.8 |
| 1994 | 199 | 35.4 |
| 1995 | 205 | 36.0 |
| 1996 | 207 | 35.8 |
| 1997 | 215 | 36.8 |
| 1998 | 223 | 37.6 |
| 1999 | 229 | 38.1 |
| 2000 | 232 | 38.2 |
| 2001 (prel) | 237 | 38.6 |

Source: FAO, FAOSTAT Statistics Database, at <apps.fao.org>, updated 7 November 2001.

# Cocoa Production Jumps

*Kathleen Huvane*

Global cocoa production in 2000 exceeded 3.2 million tons, a 10.5-percent increase from 1999 levels.[1] (See Figure 1.) Production expanded nearly threefold between 1961 and 2000.[2] And over the past century, as chocolate has become a staple rather than a luxury item in wealthy countries, production increased 24-fold.[3]

Although more than 50 nations grow cocoa, the top five producers account for over 70 percent of the total crop.[4] (See Figure 2.) Land area under cocoa cultivation increased 67 percent between 1961 and 2000, but major producing nations have scarce land resources left.[5] The economies of many producing countries hinge upon the cocoa trade. Côte d'Ivoire and Ghana, which grow three fifths of the world's cocoa, each rely on the crop for more than 20 percent of their export revenues.[6]

Falling prices in the 1990s caused Malaysian farmers to shift from cocoa to other crops like palm oil.[7] And Nigeria's cocoa industry is still rebounding from the 1970s petroleum boom that reduced the relative profitability of this crop.[8] Cocoa prices in 2000 reached record lows: three times lower than in 1960, and four times below the price in 1980.[9]

Development of the organic chocolate industry, which represents 1 percent of the chocolate market, provides an alternative for farmers seeking a greater share of the profits. Though the organic market is small, it has grown by 400 percent since 1998, and is expected to expand another 60 percent by 2002.[10]

Cacao trees grow best in humid tropical forests situated within 10 degrees of the equator.[11] As the trees age, productivity decreases, while vulnerability to pests and disease increases. Cocoa cultivated under full sun, as is two thirds of Côte d'Ivoire's crop, yields bumper crops initially, but returns diminish as soil moisture and fertility decline.[12]

Seeds of the cacao tree are ground into cocoa liquor, and separated into cocoa butter and powder. Three varieties dominate production: Criollo, Forastero, and Trinitario, a natural genetic cross.[13] The latter two account for 90 percent of production. With 40 percent fewer seeds per pod, Criollo plants have lower yields, but their superior quality fetches the highest market price.[14] Composed of 40 percent fat, 40 percent carbohydrates, and 20 percent protein, cocoa has more caffeine per liquid ounce than Pepsi-Cola.[15]

Three fourths of the 1998–99 crop was imported by Europe and the United States.[16] Most cocoa is exported whole, but producer countries are expanding their grinding operations, which accounted for 32 percent of global grindings in 2000–01.[17] Between 1996 and 1998, Côte d'Ivoire doubled its grinding capacity, capturing more profits but at the same time wedding its economy to continued cocoa production.[18]

Since chocolate may contain sugar, milk, oil, and other ingredients, chocolate consumption is not a direct measure of cocoa consumption. The average northern European eats 8.5 kilograms of chocolate annually, more than the average African eats in a lifetime.[19] Because markets in Europe and the United States are relatively saturated, producers are beginning to focus on markets in Africa, Asia, and Latin America, where four fifths of the world's population consume just one fifth of the world's cocoa.[20] (See Figure 3.)

Small landholders, who produce 90 percent of the world's cocoa, have a comparative advantage in lower labor and input costs. The estimated 15,000 children who provide forced labor to cocoa, coffee, and cotton farms in northern Côte d'Ivoire reveal the brutal tactics used by some producers to ensure profitability.[21] In December 2001, chocolate manufacturers, consumer groups, and labor advocates signed an accord addressing these labor abuses.[22]

Production of cocoa and other economically valuable non-timber forest products in the shade of the rainforest can boost local incentives for forest conservation and reduce encroachment in protected areas. Diversification leaves farmers less vulnerable to market fluctuations, diseases, and pests; reduces chemical input requirements; and provides secondary habitat and corridors for native forest species and seasonal migrants.[23]

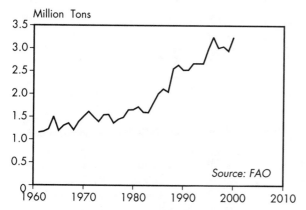

Figure 1: World Cocoa Production 1961–2000

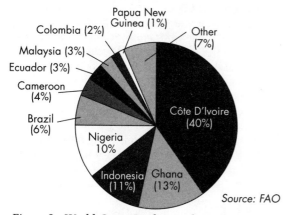

Figure 2: World Cocoa Production by Country, 2000

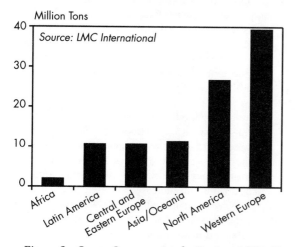

Figure 3: Cocoa Consumption by Region, 1997–98

## World Cocoa Production, 1961–2000

| Year | Production (million tons) |
|------|------|
| 1961 | 1.2 |
| 1965 | 1.2 |
| 1970 | 1.5 |
| 1971 | 1.6 |
| 1972 | 1.5 |
| 1973 | 1.4 |
| 1974 | 1.5 |
| 1975 | 1.5 |
| 1976 | 1.4 |
| 1977 | 1.4 |
| 1978 | 1.5 |
| 1979 | 1.6 |
| 1980 | 1.7 |
| 1981 | 1.7 |
| 1982 | 1.6 |
| 1983 | 1.6 |
| 1984 | 1.8 |
| 1985 | 2.0 |
| 1986 | 2.1 |
| 1987 | 2.0 |
| 1988 | 2.6 |
| 1989 | 2.6 |
| 1990 | 2.5 |
| 1991 | 2.5 |
| 1992 | 2.7 |
| 1993 | 2.7 |
| 1994 | 2.7 |
| 1995 | 3.0 |
| 1996 | 3.2 |
| 1997 | 3.0 |
| 1998 | 3.0 |
| 1999 | 3.0 |
| 2000 | 3.2 |

Sources: FAO, FAOSTAT Statistics Database, at <apps.fao.org>, updated 7 November 2001.

# Sugar and Sweetener Use Grows

*Erik Assadourian*

The consumption of sugar and other sweeteners, which are added to foods to enhance flavor, reached an estimated 157 million tons in 2001, more than 2.5 times the figure in 1961.[1] (See Figure 1.) Global per capita consumption rose from 194 calories per day in 1961 to 245 calories in 2001.[2] (See Figure 2.)

The overwhelming majority of sweetener is sugar (sucrose), derived from sugarcane and sugar beets, which contributes almost 90 percent of the sweetener supply.[3] India and Brazil, the two largest global sugar producers, produced more than a quarter of the world's sugar supply (36 million tons) in 2001.[4]

At 11.7 million tons, the next largest source of sweetener is high-fructose syrups (HFS), which are primarily produced from corn and used mostly to sweeten soft drinks.[5] HFS accounts for 7 percent of the global sweetener supply, about three quarters of which is consumed in the United States.[6] Other sweeteners include honey, maple syrup, sugar alcohols, and fruit-derived sugars, as well as high-intensity (artificial) sweeteners like saccharin and aspartame.

Link: p. 140

Worldwide, consumption of sugar increased at a modest 1 percent in 2001. Some of the fastest growth occurred in China, where it grew by 4 percent.[7] Globally, consumption of high-fructose syrup grew more rapidly, increasing 2.9 percent in 2001.[8] Over the last 10 years, HFS consumption has increased 50 percent while sugar consumption grew by 22 percent.[9]

Even faster growth has been seen in the high-intensity sweetener category. In 1999, consumption of these totaled 59,100 tons, more than a 10-fold increase since 1966.[10] As high-intensity sweeteners are anywhere from 30 to 600 times sweeter than sucrose, consumption at this level was the equivalent of using an additional 10.8 million tons of sugar.[11]

High-intensity sweeteners are essentially non-caloric, making them popular in diet beverages and foods.[12] Unlike all other sweeteners, most of these are produced not from plants but from petrochemicals. The debate continues about whether these products are harmful. The United States retracted its carcinogen warning for saccharin in 2000, while Canada has banned saccharin usage in food products since 1978.[13]

The largest consumers of sugar and sweeteners are India and the United States, having used 30 percent of the total—46 million tons—in 1999. China also used a significant amount, at 9 million tons. Considering consumption per capita, however, the United States is by far the leader—using almost three times as many sweeteners as India and 10 times as many as China.[14] (See Figure 3.) Americans on average consumed 686 calories of sweeteners a day in 1999—more than a quarter of the recommended 2,250-calorie diet.[15]

Because sweeteners are just empty calories, containing no vitamins or minerals, the World Health Organization considers them an unnecessary part of the diet.[16] Yet sweetener consumption is growing, especially in the developing world, where it has jumped 61 percent since 1961.[17] In China, per capita consumption during this period has more than tripled.[18] This growth is being pushed along by the falling costs of processed foods, growing income, heavy marketing of high-sugar foods, and urbanization, all of which are associated with eating more sweets.

Diets high in added sugars can contribute to high rates of tooth decay, especially in the absence of preventative dental care.[19] Further, as refined foods are introduced into new areas of the world, the cavity-causing effects of sugars are exacerbated by the reduction in consumption of more fibrous foods that help to inhibit decay.[20]

Sugar and sweeteners often squeeze more nutritious foods out of the diet. While Americans on average eat almost three times as much sweeteners as the recommended maximum, they eat only a third to two thirds as much fruit as they should.[21] Yet when other foods are not displaced, increased sweetener consumption can contribute to increases in obesity, which has been linked to diabetes, certain cancers, and heart disease.[22]

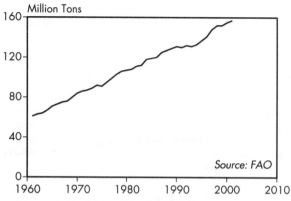

Figure 1: World Sugar and Sweetener Consumption, 1961–2001

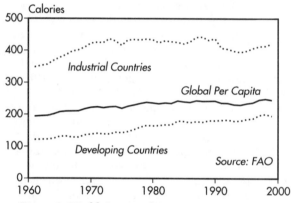

Figure 2: World Sugar and Sweetener Consumption Per Person, 1961–2000

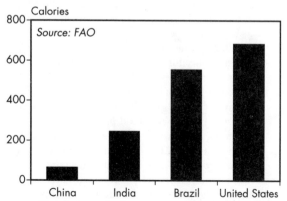

Figure 3: Daily Consumption of Sweeteners Per Person, Selected Countries, 1999

## World Sugar and Sweetener Consumption, 1961–2001

| Year | Consumption (million tons) |
|---|---|
| 1961 | 61 |
| 1965 | 71 |
| 1970 | 84 |
| 1971 | 86 |
| 1972 | 87 |
| 1973 | 89 |
| 1974 | 92 |
| 1975 | 91 |
| 1976 | 95 |
| 1977 | 99 |
| 1978 | 103 |
| 1979 | 106 |
| 1980 | 107 |
| 1981 | 108 |
| 1982 | 111 |
| 1983 | 112 |
| 1984 | 118 |
| 1985 | 119 |
| 1986 | 120 |
| 1987 | 125 |
| 1988 | 127 |
| 1989 | 129 |
| 1990 | 131 |
| 1991 | 130 |
| 1992 | 132 |
| 1993 | 131 |
| 1994 | 133 |
| 1995 | 137 |
| 1996 | 141 |
| 1997 | 148 |
| 1998 | 152 |
| 1999 | 152 |
| 2000 (prel) | 155 |
| 2001 (prel) | 157 |

Sources: FAO, *FAOSTAT Statistics Database,* at <apps.fao.org>, updated 7 November 2001; USDA, *Production Supply, and Distribution,* electronic database, December 2001.

# Irrigated Area Rises

*Janet Larsen*

In 1999, the latest year for which global figures are available, world irrigated area rose by 3 million hectares to 274 million hectares—a gain of 1.1 percent.[1] (See Figure 1.) Since peaking in 1978, irrigated land per person has declined to around 0.046 hectares.[2] (See Figure 2.)

Asia, with an increase of 1.7 percent, is responsible for the worldwide irrigation expansion in 1999.[3] This continent holds 70 percent of total irrigated area.[4] (See Figure 3). China and India claim 54 million and 59 million irrigated hectares respectively—41 percent of the total.[5]

Links: pp. 26, 102, 134

Since 1995, irrigated area in other parts of the world has remained steady, or, as in Europe and Oceania, has declined.[6] Irrigation expansion has largely bypassed Africa: just 6 percent of the continent's farmland is irrigated, up from 5 percent in 1961.[7]

The crop yield on irrigated lands is often twice that of rain-fed lands because individual plants grow better with a controlled water supply and because two or three harvests may be reaped from the same plot each year. The 274 million hectares under irrigation represent only 18 percent of farmland worldwide, but they produce some 40 percent of global agricultural goods and 60 percent of world grain supply.[8]

Some 2,500 cubic kilometers of water were applied to farmland in 1999, approaching 70 percent of all fresh water withdrawn by humans.[9] When water supplies dwindle, however, economics tends to favor industry over agriculture and in many parts of the world, water is diverted away from the field. In the last half-century, agricultural water consumption doubled but industrial consumption jumped sixfold.[10]

China, India, and the United States contain half of the world's irrigated area and produce almost half the grain supply, yet water supplies in each country show signs of depletion.[11] The water table under the North China Plain, which produces 25 percent of China's grain harvest, drops 1.5 meters annually.[12] Beneath the Punjab, India's breadbasket, the water table is falling a half-meter each year.[13] Since 1978, farmers in the southern Great Plains of the United States have cut back over 1 million hectares once watered from the Ogallala aquifer.[14] The country faces further losses if the Ogallala, which supports one fifth of U.S. irrigated land, continues to be depleted at the brisk rate of some 12 billion cubic meters a year.[15]

Worldwide tallies of irrigation area do not necessarily account for the conversion of irrigated land to other uses or the abandonment of land because of water scarcity or environmental damage. Salinization, which occurs when water evaporates from upper soil layers, leaving behind excess salts, inhibits production on one out of every five hectares of irrigated land worldwide, reducing the income of the world's farmers by more than $11 billion.[16]

Global irrigation efficiency, the ratio of water actually used by plants to the amount of water extracted, now averages only 43 percent, largely because 90 percent of the land that is artificially watered is under highly inefficient flood and furrow irrigation.[17] Improved irrigation efficiency can raise both land and water productivity.

Low-pressure and low-energy precision application sprinkler systems in the U.S. Texas High Plains, for example, at efficiencies of 80–95 percent, have produced water savings of 25–37 percent over conventional furrow systems.[18] Drip irrigation, used on an estimated 2.8 million hectares worldwide, could more than halve water use while raising yields anywhere from 20 to 90 percent.[19] Because they deliver water directly to plant roots, drip irrigation systems can have application efficiencies as high as 95 percent.[20]

Though traditionally viewed as costly and suitable only for large commercial farms, new affordable small-scale drip irrigation schemes have the potential to boost annual income for the world's rural poor by some $3 billion annually while improving food production and reducing hunger in drought-prone areas.[21] In both India and China, drip irrigation could be expanded over some 10 million hectares.[22] With water for irrigation expected to be increasingly scarce in the future, the importance of water-efficient technologies and farming practices will grow.

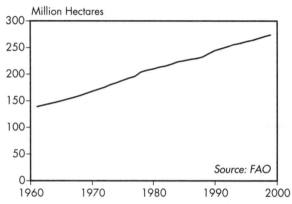

Figure 1: World Irrigated Area, 1961–99

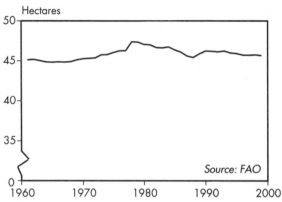

Figure 2: World Irrigated Area, Per Thousand
People, 1961–99

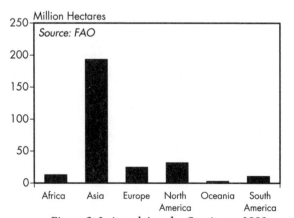

Figure 3: Irrigated Area by Continent, 1999

## World Irrigated Area and Irrigated Area Per Thousand People, 1961–99

| Year | Total | Area Per Thousand People |
|------|-------|--------------------------|
|      | (mill. hectares) | (hectares) |
| 1961 | 139 | 45.1 |
| 1965 | 150 | 44.8 |
| 1970 | 168 | 45.3 |
| 1971 | 172 | 45.3 |
| 1972 | 175 | 45.4 |
| 1973 | 180 | 45.8 |
| 1974 | 184 | 45.8 |
| 1975 | 188 | 46.0 |
| 1976 | 192 | 46.3 |
| 1977 | 196 | 46.3 |
| 1978 | 204 | 47.4 |
| 1979 | 207 | 47.3 |
| 1980 | 210 | 47.1 |
| 1981 | 213 | 47.0 |
| 1982 | 215 | 46.7 |
| 1983 | 219 | 46.6 |
| 1984 | 223 | 46.7 |
| 1985 | 225 | 46.4 |
| 1986 | 228 | 46.1 |
| 1987 | 229 | 45.6 |
| 1988 | 232 | 45.4 |
| 1989 | 238 | 45.9 |
| 1990 | 244 | 46.2 |
| 1991 | 248 | 46.2 |
| 1992 | 251 | 46.1 |
| 1993 | 256 | 46.2 |
| 1994 | 258 | 46.0 |
| 1995 | 261 | 45.9 |
| 1996 | 264 | 45.7 |
| 1997 | 267 | 45.7 |
| 1998 | 271 | 45.8 |
| 1999 | 274 | 45.7 |

Source: FAO, "Irrigation" and "Land Use,"
FAOSTAT Statistics Database, at
<apps.fao.org>, updated 10 July 2001.

# Energy Trends

Fossil Fuel Use Inches Up

Nuclear Power Up Slightly

Wind Energy Surges

Solar Cell Use Rises Quickly

Compact Fluorescents Set Record

# Fossil Fuel Use Inches Up

*Seth Dunn*

World consumption of coal, oil, and natural gas rose by 1.3 percent in 2001, to 7,956 million tons of oil equivalent, according to a preliminary estimate based on industry and government sources.[1] (See Figure 1.) Since 1950, fossil fuel use has increased by more than fourfold.[2]

Global oil consumption grew by 0.2 percent, to 3,511 million tons of oil equivalent, based on preliminary statistics from the International Energy Agency (IEA).[3] (See Figure 2.) In the United States, which accounts for 26 percent of world oil use, consumption stayed level.[4] It fell by 0.2 percent in Europe, but rose by 1.9 percent in China and declined by 0.2 percent in Asia as a whole.[5] Oil use rose the most in the former Soviet bloc and the Middle East, by 2.1 and 3.4 percent, respectively.[6] Africa registered a 1.3-percent increase in oil consumption, while Latin America logged a 1.2-percent decline.[7]

*Links*: pp. 52, 68, 132

Natural gas consumption rose by 3.2 percent to 2,233 million tons of oil equivalent.[8] The United States, with 27 percent of global natural gas use, saw a 1.9-percent drop.[9] Among industrial nations as a whole, however, gas consumption dipped by just 0.2 percent.[10]

Global coal use rose by 1.2 percent, to 2,212 million tons of oil equivalent.[11] In the United States, which uses 26 percent of world coal, consumption increased by 0.7 percent.[12] China, with a 22-percent share of coal use, saw a 1.1-percent rise, according to preliminary estimates.[13] This departure from several years of reported declines in Chinese coal use may, however, reflect a correction of official statistics that had understated consumption by excluding illegal coal mines from calculations.[14]

A major uncertainty in assessing future fossil fuel use trends is cost.[15] While improvements in technology and productivity are bringing down production and transportation costs, the cheapest reserves are being depleted, and new supplies must be brought over increasingly long distances—driving energy costs upward. As natural gas reserves near the market are depleted, for example, costs rise as

supplies must be shipped from further afield. At the same time, renewable energy resources, which can be harnessed at a local or regional level, are in general becoming less costly to produce—and more competitive with fossil sources.

Another uncertainty in projecting fossil fuel trajectories is price. In particular, oil prices are highly uncertain because they depend on the policies of major oil-producing countries. In late December 2001, ministers from the Organization of Petroleum-Exporting Countries (OPEC) committed to cutting crude oil supply during the first six months of 2002, shortly after five non-OPEC producers agreed to reduce their production or exports.[16]

As the IEA's *World Energy Outlook 2001* report points out, there are more than enough reserves of oil, gas, and coal to meet projected growth in energy demand through 2020.[17] But exploiting these reserves will require massive investments in energy production and transportation infrastructure, which in turn will have to be measured against the policy objectives of energy security and environmental protection. It is unclear, for example, how willing Middle East oil producers will be to exploit their low-cost reserves. Use of natural gas will depend, meanwhile, on the further development of technology and future prices.[18]

Renewable energy also poses a long-term threat to fossil fuels and has received added attention in the wake of the events of September 2001 and growing concern over climate change and energy security. If strong government backing achieves further reductions in the cost of renewables, the IEA study notes, there is "a huge potential for expanding the supply," which would over time cut significantly into coal use for power generation.[19] Beyond 2020, the IEA concludes, new technologies such as hydrogen-based fuel cells "hold out the prospect of abundant and clean energy supplies in a world largely free of climate-destabilising carbon emissions."[20]

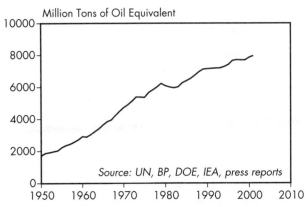

Figure 1: World Fossil Fuel Consumption, 1950–2001

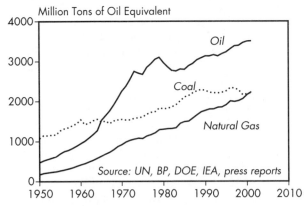

Figure 2: World Fossil Fuel Consumption, by Source, 1950–2001

## World Fossil Fuel Consumption, 1950–2001

| Year | Coal | Oil | Natural Gas |
|------|------|-----|-------------|
| | (mill. tons of oil equivalent) | | |
| 1950 | 1,074 | 470 | 171 |
| 1955 | 1,270 | 694 | 266 |
| 1960 | 1,544 | 951 | 416 |
| 1965 | 1,486 | 1,530 | 632 |
| 1970 | 1,553 | 2,254 | 924 |
| 1971 | 1,538 | 2,377 | 988 |
| 1972 | 1,540 | 2,556 | 1,032 |
| 1973 | 1,579 | 2,754 | 1,059 |
| 1974 | 1,592 | 2,710 | 1,082 |
| 1975 | 1,613 | 2,678 | 1,075 |
| 1976 | 1,681 | 2,852 | 1,138 |
| 1977 | 1,726 | 2,944 | 1,169 |
| 1978 | 1,744 | 3,055 | 1,216 |
| 1979 | 1,834 | 3,103 | 1,295 |
| 1980 | 1,814 | 2,972 | 1,304 |
| 1981 | 1,826 | 2,868 | 1,318 |
| 1982 | 1,863 | 2,776 | 1,322 |
| 1983 | 1,914 | 2,761 | 1,340 |
| 1984 | 2,011 | 2,809 | 1,451 |
| 1985 | 2,107 | 2,801 | 1,493 |
| 1986 | 2,143 | 2,893 | 1,504 |
| 1987 | 2,211 | 2,949 | 1,583 |
| 1988 | 2,261 | 3,039 | 1,663 |
| 1989 | 2,293 | 3,088 | 1,738 |
| 1990 | 2,270 | 3,136 | 1,774 |
| 1991 | 2,225 | 3,134 | 1,806 |
| 1992 | 2,211 | 3,165 | 1,810 |
| 1993 | 2,206 | 3,135 | 1,849 |
| 1994 | 2,224 | 3,192 | 1,858 |
| 1995 | 2,258 | 3,235 | 1,913 |
| 1996 | 2,342 | 3,316 | 2,005 |
| 1997 | 2,327 | 3,388 | 1,993 |
| 1998 | 2,281 | 3,398 | 2,016 |
| 1999 | 2,160 | 3,469 | 2,065 |
| 2000 | 2,186 | 3,504 | 2,164 |
| 2001 (prel) | 2,212 | 3,511 | 2,233 |

Source: Worldwatch estimates based on UN, BP, DOE, IEA, and press reports.

# Nuclear Power Up Slightly

*Nicholas Lenssen*

Between 2000 and 2001, total installed nuclear power generating capacity increased by 1,505 megawatts (0.4 percent), passing 350,000 megawatts for the first time.[1] (See Figure 1.) But since 1990, global nuclear capacity has risen just 7 percent—compared with 240-percent growth in the 1980s—an indication of nuclear power's stagnation in the past decade.[2]

Only one new reactor was grid-connected in 2001, in Russia, bringing the world's total to 436.[3] The remaining capacity increase in 2001 is due to upgrades at existing reactors, where more power was squeezed from operating units. Last year, for the first time ever, there was neither new construction started on a reactor (see Figure 2) nor any operating reactors permanently shut down.[4]

Some 26 reactors remain under active construction (with a combined capacity of 23,537 megawatts), with as many as eight of these due for completion in 2002.[5] And a total of 99 reactors (representing more than 30,000 megawatts) have been retired after an average service life of less than 18 years.[6] (See Figure 3.)

In the United States, 2001 started with industry and government talking about a "nuclear renaissance." The new administration touted nuclear power in its energy plan, and power shortages in California encouraged promoters in believing that the country would seriously consider initiating a new nuclear project for the first time since the early 1970s.[7]

The terrorist attacks of September 11th, however, quickly put a damper on these aspirations: armed troops were deployed around existing reactors, and even the International Atomic Energy Agency confessed that little could be done to protect nuclear power plants from such airborne attacks.[8]

Official or de facto moratoria remain on new nuclear power in most of Western Europe. Belgium reiterated its plan to shut down existing plants before they are 40 years old, and the German government and industry formalized an agreement to phase out existing reactors.[9]

The United Kingdom considered an energy policy that would include building new reactors, but instead chose to rely on renewable energy sources such as wind.[10] In Sweden, the coalition government moved to postpone the planned closure of a unit until 2003 due to the concern that replacement power would not yet be available.[11]

A breath of life returned to Russia's nuclear program in 2001, as economic recovery resulted in more funding. In addition to the one reactor completed in 2001, work restarted on two others, with plans calling for as many as 10 new reactors in the next decade.[12] Russia may also help Ukraine complete two reactors stalled since the 1986 Chernobyl meltdown.[13]

Japan's nuclear program continues to face local opposition as public referenda in Kariwa Village and Miyama resulted in votes against nuclear projects.[14] Another planned plant was "temporarily suspended" due to local opposition in Amori Prefecture.[15] Only four units were under construction in Japan, with two more units in pre-construction safety review.

China has the world's largest nuclear expansion effort, with 10 reactors being built to go along with its three operating units.[16] Four of the new units are likely to be grid-connected in 2002, and the country initiated work on a site for as many as four more new ones.[17] South Korea has four reactors under construction.[18] And Taiwan restarted building two units in 2001 after the government's move to scrap the plant in 2000 was declared unconstitutional.[19] But the election victory by the Progress Democrat Party in late 2001 is likely to halt the project once again.[20]

Numerous other countries—including Argentina, Brazil, India, and Romania—continue to discuss restarting stalled projects or ordering new units, but none of these discussions have yet turned into secure financing, much less cement being poured.

Indeed, in a post–September 11th world, many countries and policymakers have reason to reevaluate nuclear energy. The threat extends beyond the simple disruption of nuclear power plant operation to the trafficking of nuclear materials. On two occasions in late 2001, for example, police arrested black marketers attempting to sell weapons-grade enriched uranium in Russia and Turkey.[21]

Gigawatts

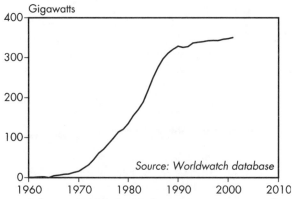

Figure 1: World Electrical Generating Capacity of
Nuclear Power Plants, 1960–2001

Gigawatts

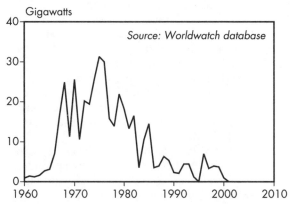

Figure 2: World Nuclear Reactor Construction Starts,
1960–2001

Gigawatts

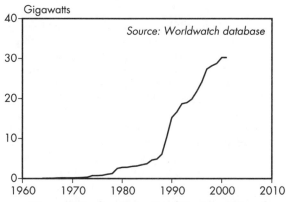

Figure 3: Nuclear Capacity of Decommissioned
Plants, 1964–2001

| World Net Installed Electrical Generating Capacity of Nuclear Power Plants, 1960–2001 | |
|---|---|
| Year | Capacity |
| | (gigawatts) |
| 1960 | 1 |
| 1965 | 5 |
| 1970 | 16 |
| 1971 | 24 |
| 1972 | 32 |
| 1973 | 45 |
| 1974 | 61 |
| 1975 | 71 |
| 1976 | 85 |
| 1977 | 99 |
| 1978 | 114 |
| 1979 | 121 |
| 1980 | 135 |
| 1981 | 155 |
| 1982 | 170 |
| 1983 | 189 |
| 1984 | 219 |
| 1985 | 250 |
| 1986 | 276 |
| 1987 | 297 |
| 1988 | 310 |
| 1989 | 320 |
| 1990 | 328 |
| 1991 | 325 |
| 1992 | 327 |
| 1993 | 336 |
| 1994 | 338 |
| 1995 | 340 |
| 1996 | 343 |
| 1997 | 343 |
| 1998 | 343 |
| 1999 | 346 |
| 2000 | 348 |
| 2001 (prel) | 351 |

Source: Worldwatch Institute database,
compiled from the IAEA and press reports.

# Wind Energy Surges
*Christopher Flavin*

Wind energy generating capacity jumped 37 percent, to approximately 24,800 megawatts at the end of 2001.[1] (See Figure 1.) The capacity addition of roughly 6,700 megawatts during the year was up sharply from the year before—and reinforces wind's position as the world's fastest growing energy source.[2] (See Figure 2.) Annual wind capacity additions are now approaching annual additions to global hydropower capacity, and are more than four times the nuclear capacity added in 2001.[3]

Europe now has over 70 percent of the world's wind capacity, thanks mainly to the strong laws encouraging its growth in Germany, Spain, and Denmark.[4] Germany strengthened its role as the world leader in 2001, with 2,600 megawatts added, taking its capacity to over 8,700 megawatts—more than one third of the world total.[5] Wind power now provides 3.5 percent of Germany's electricity, and the government has announced plans to raise that figure to at least 25 percent by 2025, while phasing out the nuclear industry, which now provides 30 percent of the country's power.[6]

Spain established a clear position as Europe's second leading wind generator in 2002, with an additional 1,100 megawatts—taking its total to 3,340 megawatts and providing an estimated 3 percent of the country's electricity.[7] Spain's wind industry is becoming an increasingly important international player, with ventures now under way in other parts of Europe, Latin America, and China. The country's leading wind company, Gamesa Eolica, linked to one of the country's leading aeronautical and industrial enterprises, was 40-percent owned by Denmark's Vestas until the end of 2001, when the Gamesa Group acquired those shares in order to be able to compete with Vestas in markets around the world.[8]

Denmark, which gets a world-leading 18 percent of its electricity from the wind, saw a sharp slowdown in its pace of growth in 2001, with just over 100 megawatts added, taking its total to 2,400 megawatts.[9] The slowdown stems from a government decision in 2000 to end the minimum purchase price requirement and introduce a new system of renewable certificate trading that has not been successfully implemented so far.[10] The situation turned even bleaker in early 2002, when a new right-wing government announced plans to dismantle the country's remaining support for wind energy.[11]

Countering the negative trend in Denmark was Italy's addition of 270 megawatts, moving it into the fourth position in Europe, with nearly 700 megawatts installed.[12] And outside Europe, India reinvigorated its wind power industry in 2001, with an added 300 megawatts, taking the national total to 1,500 megawatts installed.

The United States rejoined the wind energy big leagues in 2001, with nearly 1,700 megawatts added, a rush spurred by a federal tax credit that expired at the end of the year.[13] The country's total installed capacity remains second to Germany's, as it has since the late-1990s. The record-breaking U.S. installations were spread broadly across the country's western plains and mountains, with major projects in Texas, Kansas, and Oregon.[14] Even larger projects are planned, following congressional reinstatement of the federal wind energy tax credit in March 2002.[15]

The newest player on the wind energy scene is Brazil, which was hit hard by drought-induced power shortages in 2001, and is now turning to wind as a quick and affordable way of boosting its generating capacity. Some 4,000 megawatts of wind power projects were authorized by Brazil's federal electricity regulator, Aneel, in late 2001 and early 2002—which could make Brazil the world's fourth largest market in the next two years.[16] Much of the development is occurring in the economically deprived but wind-rich northeastern states of Rio Grande do Norte, Ceara, Pernambuco, and Bahia.[17]

The global wind power industry generated an estimated $7 billion in business in 2001, and is now attracting the interest of the world's largest energy companies, ranging from ABB to Royal Dutch Shell.[18] Another major player joined the scene in early 2002 when General Electric reached agreement with the bankrupt Enron Corp to purchase the company's wind energy business, which is the largest in North America.[19]

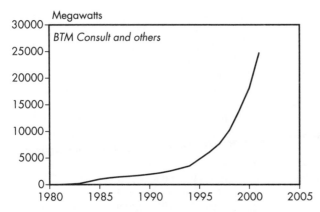

Figure 1: World Wind Energy Generating Capacity,
1980–2001

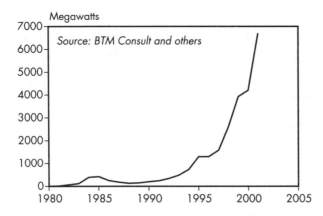

Figure 2: Annual Addition to World Wind Energy
Generating Capacity, 1980–2001

## World Wind Energy Generating Capacity, Total and Annual Addition, 1980–2001

| Year | Total | Annual Addition |
|---|---|---|
| | (megawatts) | |
| 1980 | 10 | 5 |
| 1981 | 25 | 15 |
| 1982 | 90 | 65 |
| 1983 | 210 | 120 |
| 1984 | 600 | 390 |
| 1985 | 1,020 | 420 |
| 1986 | 1,270 | 250 |
| 1987 | 1,450 | 180 |
| 1988 | 1,580 | 130 |
| 1989 | 1,730 | 150 |
| 1990 | 1,930 | 200 |
| 1991 | 2,170 | 240 |
| 1992 | 2,510 | 340 |
| 1993 | 2,990 | 480 |
| 1994 | 3,490 | 730 |
| 1995 | 4,780 | 1,290 |
| 1996 | 6,070 | 1,290 |
| 1997 | 7,640 | 1,570 |
| 1998 | 10,150 | 2,600 |
| 1999 | 13,930 | 3,920 |
| 2000 | 18,100 | 4,200 |
| 2001 (prel) | 24,800 | 6,700 |

Sources: BTM Consult, EWEA, AWEA, *Windpower Monthly*, and *New Energy*.

# Solar Cell Use Rises Quickly

*Molly O. Sheehan*

Production of photovoltaic (PV) cells, which turn sunlight into electricity, exceeded 390 megawatts in 2001, according to a survey of manufacturers.[1] The 36-percent surge made 2001 the fourth straight year of growth at or above 30 percent.[2] (See Figure 1.) The 1,140 megawatts of installed PVs in the world today have just a bit more capacity than the largest coal-fired power plant and account for less than 1 percent of global electricity.[3] But if current growth is sustained, PVs could become a globally significant power source within the next three decades.[4]

Government support in a few industrial nations has powered the PV market recently, prompting a dramatic leap in the share of PVs that supplement existing power grids. Grid-connected PVs accounted for only 14 percent of solar power installed in 1995, but by 2000, they accounted for more than 50 percent, according to a survey by Strategies Unlimited.[5]

Japan has subsidized tens of thousands of PV rooftops since 1996.[6] The government paid for 50 percent of a new solar system when it first launched the program, although by 2001 it had lowered the subsidy to 15 percent.[7] Japanese manufacturers produced just under 44 percent of the global output in 2001, keeping Japan in the lead as the world's largest PV producer.[8] (See Figure 2.) As much as 120 of the 171 megawatts of PV cells produced in Japan in 2001 were used in that country.[9]

Support for PVs is also strong in Europe, where 86.3 megawatts were produced in 2001.[10] Government initiatives helped spur the purchase of some 65 megawatts in Germany alone, and more than 20 megawatts in other European nations.[11]

Although the United States is the second largest producer of solar cells, with an output of 100.3 megawatts in 2001, most of this product is exported.[12] State and city initiatives, led by California, are starting to lower barriers for solar, however, and enlarge the market.[13] In 2000, just 12 megawatts were purchased in the United States, but that grew by 50 percent in 2001, with 10 megawatts sold in California alone.[14]

Over the last two decades, mass production and technological advances have slashed the cost of PVs, but strong demand since the mid-1990s has slowed the decline in prices.[15] Some 90 percent of PVs produced in 2001 were made from crystalline silicon, which is sliced into wafers and encased in glass panels.[16] The remaining 10 percent is cheaper but less efficient "thin-film" silicon, which can be made into flexible sheets and integrated into building materials.[17] Industry analyst Paul Maycock now quotes two factory prices for PVs: $3.50 per watt for the crystalline PVs and $2 for the less efficient thin-film variety.[18]

The solar arrays being installed in industrial nations fill an urban niche, helping cities avoid blackouts during peak air conditioning demand.[19] As PVs can be mounted directly on homes and businesses, power can be used right where it is generated, eliminating transmission losses. Without subsidies, the price of installed solar power can be several times the average retail electricity price, but it can be competitive at times of peak use.[20]

There is even greater need for off-grid PVs in the developing world, where some 1.7 billion people live without access to electricity that can help boost education, health, and income by powering water pumps, refrigeration for vaccines, computers, and communications.[21] Since 1991, up to $520 million in loans have been pledged by the World Bank Group to support this market; so far, such support has resulted in the installation of an estimated 500,000 off-grid, residential solar systems.[22]

The total number of solar systems in the developing world is likely much higher, as nonprofit groups and private entrepreneurs have been helping to devise financing and credit schemes so that businesses and consumers can overcome the high up-front cost of PVs.[23] By offering loans, partnering with local microfinance partners, or selling a "fee-for-service" package, the Solar Electric Light Company has sold more than 16,000 solar home systems in India, Sri Lanka, and Viet Nam since it was launched in 1997.[24]

Megawatts

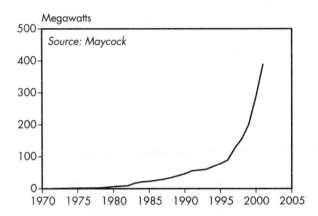

Figure 1: World Photovoltaic Production, 1971–2001

Megawatts

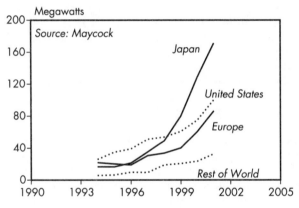

Figure 2: Photovoltaic Production by Country
or Region, 1994–2001

## World Photovoltaic Production, 1971–2001

| Year | Production (megawatts) |
|---|---|
| 1971 | 0.1 |
| 1975 | 1.8 |
| 1976 | 2.0 |
| 1977 | 2.2 |
| 1978 | 2.5 |
| 1979 | 4 |
| 1980 | 7 |
| 1981 | 8 |
| 1982 | 9 |
| 1983 | 17 |
| 1984 | 22 |
| 1985 | 23 |
| 1986 | 26 |
| 1987 | 29 |
| 1988 | 34 |
| 1989 | 40 |
| 1990 | 46 |
| 1991 | 55 |
| 1992 | 58 |
| 1993 | 60 |
| 1994 | 69 |
| 1995 | 79 |
| 1996 | 89 |
| 1997 | 126 |
| 1998 | 153 |
| 1999 | 201 |
| 2000 | 288 |
| 2001 (prel) | 391 |

Source: Paul Maycock, PV News, various issues.

*Michael Scholand*

Global sales of energy-efficient compact fluorescent lamps (CFLs) grew by 15 percent in 2001, achieving record levels of over 600 million units.[1] (See Figures 1 and 2.) CFLs, a technology with nearly two decades of commercialization, are a tiny version of the common 4-foot fluorescent tubes. Compared with incandescent light bulbs they are designed to replace, quality CFLs last about 10 times longer and use just one quarter of the electricity while providing the same amount of light.[2]

Between 1988 and 2001, CFL sales increased more than 13-fold.[3] There are an estimated 1.8 billion CFLs in operation today, consuming 27,000 megawatts of electricity—much less than the 109,000 megawatts that would be required to operate the same number of incandescent lamps.[4] The electricity these CFLs are saving is equivalent to that produced by nearly 40 medium-sized coal-fired power plants.[5]

Link: p. 132

Avoided electricity generation translates into pollution reduction. In North America, the 316 million CFLs in use at the start of 2002 will save 4.8 million tons of carbon and 94,000 tons of sulfur dioxide emissions during the year.[6] CFLs also reduce energy bills: in Thailand, consumers pay about 300 baht ($6.70) for a high-quality CFL that, if lit four hours a day, offers a payback on the additional first cost in just 1.5 years.[7] Looking at bulb replacement and electricity savings over the 10,000-hour life of the lamp, a CFL has a net present value of over 1,000 baht—more than three times what it cost.[8]

Recognizing a great opportunity, China launched a three-year Green Lights program in January 1997 to expand their efficient-lighting market and improve production quality.[9] The program covers education, certification, labeling, demonstrations, and technical assistance.[10] China's CFL industry expanded, fueled by this government support and a robust domestic market that grew by over 350 percent in the last six years.[11] Today, China manufactures more than 80 percent of the world's CFL supply.[12] Growing sales volumes stimulated competition and innovation, reducing prices and improving quality—trends that continue today.[13]

Recognizing the economic benefits of CFLs, the International Finance Corporation (IFC) launched the Efficient Lighting Initiative (ELI) in 1999 with support from the Global Environment Facility. Now in its third year, ELI is expanding the market for efficient lighting in seven countries: Argentina, the Czech Republic, Hungary, Latvia, Peru, the Philippines, and South Africa.[14] Russell Sturm, program manager at the IFC, indicates that accelerating market adoption of efficient lighting will not only reduce household expenditures, it will also help countries meet their energy needs more cost-effectively and reduce greenhouse gas emissions for less than $5 per ton.[15]

In South Africa, ELI has succeeded in raising awareness and gaining market acceptance for this previously unknown technology. Barry Bredenkamp of Bonesa, the organization coordinating ELI in South Africa, reported that "after only two years, our program achieved an estimated 59 percent growth in CFL sales in the last year, increasing annual sales to over 4 million lamps. We have targeted customers across income groups, including lower-income households that often install a CFL as their first electric light source."[16]

In Peru, South Africa, Argentina, and the Philippines, which have seen an influx of low-cost, low-quality CFLs, ELI provides a labeling scheme to certify the lamps, protecting consumers against counterfeit or inferior products. This kind of quality assurance is particularly crucial for the first-time buyer, who could reject the technology outright after a negative experience.[17]

While CFL bulbs have many environmental and economic advantages, like all fluorescent lamps they do contain a small amount of mercury. Recently, manufacturers have succeeded in reducing the mercury content to less than 5 milligrams per CFL, or about 1 percent as much as a household thermometer.[18] Moreover, in the United States energy savings from CFLs cut environmental mercury emissions by reducing electricity produced from coal-burning power stations.[19] Consumers can also recycle their CFLs, thereby eliminating any environmental contamination.

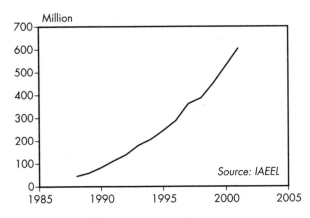

Figure 1: World Sales of Compact Fluorescent
Lamps, 1988–2001

| World Sales of Compact Fluorescent Lamps, 1988–2001 | |
| --- | --- |
| Year | Units (million) |
| 1988 | 45 |
| 1989 | 59 |
| 1990 | 83 |
| 1991 | 112 |
| 1992 | 138 |
| 1993 | 179 |
| 1994 | 206 |
| 1995 | 245 |
| 1996 | 288 |
| 1997 | 362 |
| 1998 | 387 |
| 1999 | 452 |
| 2000 | 528 |
| 2001 (prel) | 606 |

*Source:* Nils Borg, IAEEL, e-mails to World-watch; 1988–89 from Evan Mills, Lawrence Berkeley Laboratory.

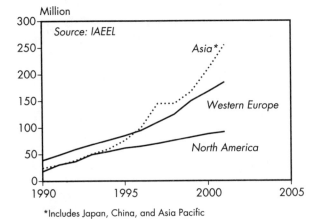

*Includes Japan, China, and Asia Pacific

Figure 2: World Sales of Compact Fluorescent Lamps,
Selected Regions, 1990–2001

# Atmospheric Trends

Global Temperature Close to a Record

Carbon Emissions Reach New High

CFC Use Declining

# Global Temperature Close to a Record

*Seth Dunn*

Global surface air temperatures rose to 14.43 degrees Celsius in 2001, based on land and ocean measurements dating back to 1880 from NASA's Goddard Institute for Space Studies (GISS). (See Figure 1.)[1] Another GISS dataset, based only on land measurements but extending back to 1867, showed similar results. (See Figure 2.)[2] Both indicate that 2001 was the second warmest year on record—a finding supported by datasets from the U.S. National Oceanic and Atmospheric Administration and the U.K. Hadley Centre for Climate Prediction.[3] The warmest year thus far was 1998.[4]

Regional surface patterns reflected above-average temperature conditions, though large parts of the tropical and north Pacific were cooler than average.[5] Canada has now had 18 straight seasons of above-average temperatures.[6] October 2001 was the warmest October in the 343-year Central England temperature series.[7] Some regions experienced unusually cool weather, however.[8] Russia registered more than 100 deaths from hypothermia from low temperatures during the 2000–01 winter.[9] And northern India saw more than 130 deaths from extreme cold in January 2002.[10]

Link: p. 52

Several areas experienced above- or below-average rainfall.[11] The period from March 1999 to March 2001 was the wettest 24-month period in the 236-year time series for England and Wales.[12] India experienced its second lowest total of winter precipitation in early 2001, and a drier-than-normal summer monsoon season that exacerbated water shortages in some areas.[13] Several countries, such as Australia and Zambia, experienced a mix of both wetter and drier weather in various locations.[14]

The year saw an above-average number of hurricanes and tropical storms in the north Atlantic basin, with 15 named storms—5 more than the long-term average.[15] Tropical Storm Allison caused the most extensive flooding in the United States ever associated with a tropical storm.[16] Hurricane Michelle severely affected the coffee crop in Jamaica and was the strongest hurricane to make landfall in Cuba since 1952.[17] In the western Pacific, Typhoon Chebi reached sustained winds of close to 160 kilometers an hour, killing at least 79 people.[18]

Unusual flood events were reported.[19] Mozambique and Zambia experienced as many as 200 deaths from heavy rainfall that ruined crops and left hundreds homeless.[20] Hungary's rain-swollen Tisza River reached its highest level since 1888.[21] In Siberia, rainfall and accelerated snowmelt caused ice-jammed rivers to overflow, destroying or damaging the homes of more than 300,000 people.[22] Viet Nam's Mekong Delta region saw several hundred deaths from October flooding.[23] Heavy rains in West Africa affected nearly 70,000 people and submerged 17,000 hectares of agricultural land.[24] And hundreds were killed in Algiers from Algeria's worst flooding in almost 40 years.[25]

Drought affected many areas. The region encompassing Iran, Afghanistan, and Pakistan continued to suffer from a devastating drought that began in 1998, with a wet season more than 45 percent below average precipitation.[26] This lack of rainfall has stressed both water supplies and agriculture, directly affecting more than 60 million people.[27] The region was also subject to periods of extreme heat, one of which caused many deaths in Pakistan in early May.[28] Drought persisted in the Greater Horn region of Africa; in Brazil, exacerbating the nation's hydropower supply shortage; and in northern China, the Korean peninsula, and Japan in the first half of 2001.[29] Winter precipitation deficits affected the western United States, and Canada reported drought in many regions from coast to coast.[30]

These climatic phenomena are likely to become more frequent and intense as surface temperatures rise, according to the latest assessment of the Intergovernmental Panel on Climate Change.[31] Concluding that "there is new and stronger evidence that most of the warming observed over the last 50 years is attributable to human activities," the panel projects that average global surface temperature will increase by 1.4–5.8 degrees Celsius between 1990 and 2100.[32] The actual temperature rise will be largely determined by future trends in greenhouse gas emissions.

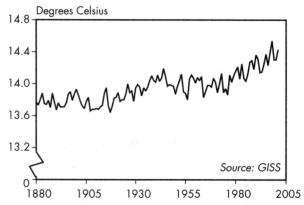

Figure 1: Global Average Temperature at Earth's
Surface, 1880–2001

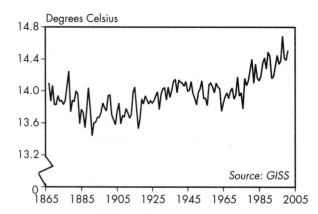

Figure 2: Global Average Temperature at Earth's Surface
(Land-Based Series), 1867–2001

## Global Average Temperature, 1950–2001

| Year | Temperature (degrees Celsius) |
|---|---|
| 1950 | 13.87 |
| 1955 | 13.88 |
| 1960 | 14.01 |
| 1965 | 13.90 |
| 1970 | 14.02 |
| 1971 | 13.89 |
| 1972 | 14.00 |
| 1973 | 14.13 |
| 1974 | 13.89 |
| 1975 | 13.94 |
| 1976 | 13.86 |
| 1977 | 14.11 |
| 1978 | 14.02 |
| 1979 | 14.10 |
| 1980 | 14.16 |
| 1981 | 14.21 |
| 1982 | 14.06 |
| 1983 | 14.25 |
| 1984 | 14.07 |
| 1985 | 14.03 |
| 1986 | 14.12 |
| 1987 | 14.27 |
| 1988 | 14.29 |
| 1989 | 14.18 |
| 1990 | 14.36 |
| 1991 | 14.31 |
| 1992 | 14.14 |
| 1993 | 14.15 |
| 1994 | 14.25 |
| 1995 | 14.37 |
| 1996 | 14.23 |
| 1997 | 14.39 |
| 1998 | 14.54 |
| 1999 | 14.30 |
| 2000 | 14.30 |
| 2001 (prel) | 14.43 |

Source: Surface Air Temperature Analysis,
Goddard Institute for Space Studies,
25 January 2002.

Global emissions of carbon from fossil fuel combustion increased by 1.1 percent in 2001, reaching a new high of 6.55 billion tons.[1] (See Figure 1.) This was the second consecutive record-setting year, and the eighth annual record since 1990. Annual carbon emissions have now more than quadrupled since 1950.[2]

Behind the global trend, national and regional emissions patterns vary widely. (National data are available only through 2000.) The United States, which accounts for 24 percent of the global total, registered an 18.1-percent increase between 1990 and 2000.[3] In contrast, emissions in the European Union (EU) over this period fell by 1.8 percent, owing mainly to declines in Germany and the United Kingdom of 19 and 5 percent, respectively.[4] The steepest drops in carbon emissions occurred in former Eastern bloc nations. Russia, for example, had a 30.7-percent decline.[5]

*Links: pp. 38, 50, 114*

Collectively, industrial and former Eastern bloc nations saw a 1.7-percent drop in carbon emissions between 1990 and 2000.[6] This compares with the commitment of these nations, under the 1997 Kyoto Protocol, to reduce emissions of carbon dioxide and other greenhouse gases by 5.2 percent between 1990 and 2010.[7]

Carbon emissions trends among the larger developing nations were generally upward, although starting from a smaller base. China's carbon emissions grew by 7.7 percent over the decade, while those of India increased 67 percent.[8] However, per capita emissions in China and India—at 0.68 and 0.3 tons—are well below the global average of 1.1 tons, and roughly one seventh and one fourteenth that of the U.S. average.[9]

The carbon intensity of the world economy continued its gradual decline, falling to 150 tons per million dollars of economic output.[10] (See Figure 2.) This represents a 40-percent decline in carbon intensity since 1950, with half of the decline occurring since 1982.[11] This "decarbonization" trend needs to be accelerated, however, to achieve a 60–80 percent reduction in carbon emissions during this century—which is what scientists believe is necessary to stabilize atmospheric concentrations of carbon dioxide ($CO_2$) below a doubling of pre-industrial levels.[12]

Atmospheric $CO_2$ levels rose to 370.9 parts per million volume (ppmv) in 2001, according to measurements from the Mauna Loa Observatory in Hawaii, part of a record dating back to 1957.[13] (See Figure 3.) The annual increase of 1.49 ppmv, up from 1.11 ppmv the previous year, suggests the possible onset of another El Niño—a climatic phenomenon related to surface warming of the Pacific Ocean.[14] The previous El Niño, in 1997–99, saw annual rises in $CO_2$ levels of 2.87 and 1.66 ppmv.[15]

Prospects for reducing carbon emissions improved in late 2001 when more than 170 nations finalized the rules for the Kyoto Protocol at talks in Marrakesh, Morocco.[16] For the protocol to enter into force, it must be ratified by 55 countries representing 55 percent of the 1990 emissions of industrial and former Eastern bloc nations—called Annex I nations under the original Framework Convention on Climate Change.[17] As of March 2002, 49 parties had ratified or acceded to the protocol, but they represented only 2.4 percent of Annex I emissions—as the Czech Republic and Romania are the only Annex I ratifiers thus far.[18]

The United States, with 36 percent of the Annex I share, withdrew from the Kyoto negotiations in 2001, and in March 2002 announced a set of voluntary measures and incentives for energy efficiency and renewable energy.[19] These steps represent more a continuation of previous policy than a new initiative, however, and are unlikely to restrain U.S. emissions growth.

The U.S. absence implies that, for the Kyoto Protocol to become law, the EU, Russia, Japan, Australia, and Canada must all ratify the pact.[20] In March 2002, EU environment ministers agreed to ratify the protocol by June and directed member state parliaments to ratify the treaty under national law; they also called on Japan and Russia to follow their lead.[21] The governments of Denmark, France, Luxembourg, and Portugal have already approved ratification of Kyoto; other EU members stated their intent to do so by mid-June.[22]

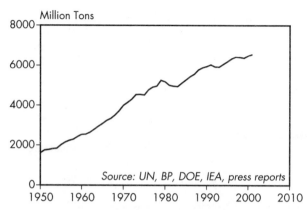

Figure 1: World Carbon Emissions from
Fossil Fuel Burning, 1950–2001

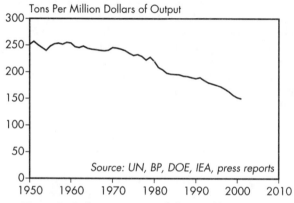

Figure 2: Carbon Intensity of the World Economy,
1950–2001

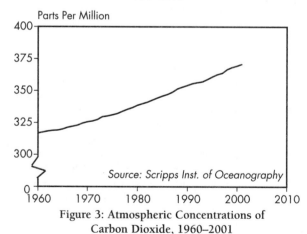

Figure 3: Atmospheric Concentrations of
Carbon Dioxide, 1960–2001

### World Carbon Emissions from Fossil Fuel Burning, 1950–2001, and Atmospheric Concentrations of Carbon Dioxide, 1960–2001

| Year | Emissions (mill. tons of carbon) | Carbon Dioxide (parts per mill.) |
|------|------|------|
| 1950 | 1,612 | n.a. |
| 1955 | 2,013 | n.a. |
| 1960 | 2,535 | 316.7 |
| 1965 | 3,087 | 319.9 |
| 1970 | 3,997 | 325.5 |
| 1971 | 4,143 | 326.2 |
| 1972 | 4,305 | 327.3 |
| 1973 | 4,538 | 329.5 |
| 1974 | 4,545 | 330.1 |
| 1975 | 4,518 | 331.0 |
| 1976 | 4,776 | 332.0 |
| 1977 | 4,910 | 333.7 |
| 1978 | 4,962 | 335.3 |
| 1979 | 5,249 | 336.7 |
| 1980 | 5,177 | 338.5 |
| 1981 | 5,004 | 339.8 |
| 1982 | 4,959 | 341.0 |
| 1983 | 4,942 | 342.6 |
| 1984 | 5,113 | 344.2 |
| 1985 | 5,274 | 345.7 |
| 1986 | 5,436 | 347.0 |
| 1987 | 5,558 | 348.7 |
| 1988 | 5,774 | 351.3 |
| 1989 | 5,879 | 352.7 |
| 1990 | 5,939 | 354.0 |
| 1991 | 6,025 | 355.5 |
| 1992 | 5,922 | 356.4 |
| 1993 | 5,914 | 357.0 |
| 1994 | 6,050 | 358.9 |
| 1995 | 6,182 | 360.9 |
| 1996 | 6,327 | 362.6 |
| 1997 | 6,419 | 363.8 |
| 1998 | 6,401 | 366.6 |
| 1999 | 6,366 | 368.3 |
| 2000 | 6,480 | 369.4 |
| 2001 (prel) | 6,553 | 370.9 |

*Source:* Worldwatch estimates based on UN, BP, DOE, IEA, and press reports.

# CFC Use Declining

*Molly O. Sheehan*

Global production of chlorofluorocarbons (CFCs), which harm Earth's protective ozone layer, fell by less than 1 percent between 1998 and 1999, the most recent year for which relatively complete data are available.[1] (See Figure 1.) CFCs were once widely used as coolants, aerosol propellants, and industrial solvents, and in foam insulation. A 1987 treaty to protect the ozone layer initiated dramatic declines in CFC output, which is now many times below peak production years, the late 1980s.[2]

China, India, and Russia produced the most CFCs in 1999.[3] (See Figure 2.) Developing nations are the largest producers because the 1987 Montreal Protocol and its amendments banned CFC production in industrial nations as of 1996, except for a small volume for export to developing countries or for essential uses, such as asthma inhalers.[4] One of the largest manufacturing plants in the industrial world, in the Netherlands, will close at the end of 2005.[5] All CFC production ceased in Russia in December 2000.[6] The Montreal Protocol requires developing countries to phase CFCs out by 2010. Many nations, including China and India, are receiving assistance from the treaty's Multilateral Fund to make this transition.[7]

Link: p. 114

Many CFCs were initially replaced by hydrochlorofluorocarbons (HCFCs), which are now being supplanted by hydrofluorocarbons because HCFCs harm the ozone layer too, albeit to a lesser extent.[8] All fluorocarbons, however, are potent greenhouse gases, so some CFC alternatives bypass this family of chemicals altogether. "Greenfreeze" refrigerators, for example, use hydrocarbons instead of fluorocarbons for coolant and insulating foam. Some 55 million Greenfreeze refrigerators dominate markets in Western Europe.[9] Prodded by Greenpeace, three major Japanese companies announced in late 2001 they would produce this type of refrigerator too.[10]

The government of Canada estimates that only 5 percent of vehicle air conditioners there still used CFCs as of mid-2001.[11] In contrast, a U.S. survey in 2001 found that building owners were only nearing the halfway point in replacing chillers that use CFCs—and that it would take at least until 2010 to complete the conversion, much longer than expected.[12] (The CFCs still used in this equipment are either recycled or obtained illegally.)

Indeed, fed by production in developing countries, a black market is thriving in industrial nations where CFC-using appliances are still in use.[13] Illegal exports from India and China have been growing.[14] Since 1995, when the United States launched a national enforcement initiative, more than 100 people have been convicted of smuggling CFCs into the country.[15]

Another threat to the ozone layer is that some chemicals originally touted as replacements to CFCs are not as benign as scientists hoped. A scientific panel advised the treaty secretariat in October 2001 to ban n-propyl bromide, hexachlorobutadiene, Halon-1202, and 6-bromo-2-methoxy-napthalene.[16] "We cannot be complacent. If enough of these new chemicals are manufactured, we will delay the recovery of the ozone layer quite significantly," warned Mario Molina, who shared the 1995 chemistry Nobel Prize for his work on ozone loss.[17]

Although CFC production has declined steeply, the ozone layer has yet to recover, as these compounds take years to reach the upper atmosphere and last for decades or centuries once there. In October 2001, researchers at the U.S. National Oceanic and Atmospheric Administration said the seasonal "hole" in the ozone layer above Antarctica appeared to have stabilized for the previous three years.[18] In September 2001, satellite data showed that the geographic area covered by the ozone hole area was about the same as the year before.[19]

Scientists in late 2000 predicted that the hole in the ozone layer should begin to close within a decade, healing completely by 2050.[20] Unfortunately, some damage has already been done: skin cancer reportedly rose 66 percent between 1994 and 2001 in Punta Arena, Chile, the world's southernmost city.[21]

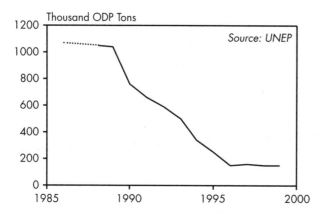

Figure 1: World CFC Production, 1986–99

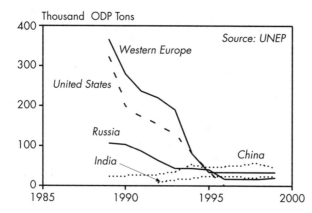

Figure 2: CFC Production by Country or Region,
1989–99

### World CFC Production, 1986–99

| Year | Production (thousand ODP tons)* |
|------|------------|
| 1986 | 1072.3 |
| 1989 | 1046.0 |
| 1990 | 764.3 |
| 1991 | 664.3 |
| 1992 | 590.8 |
| 1993 | 506.0 |
| 1994 | 338.5 |
| 1995 | 253.8 |
| 1996 | 151.6 |
| 1997 | 158.8 |
| 1998 | 146.9 |
| 1999 | 146.8 |

*These numbers reflect the volume of the major CFCs (CFC-11, CFC-12, CFC-113, CFC-114, and CFC-115) multiplied by their respective ozone-depleting potentials (ODPs). The ODP value is the ratio of a given compound's ability to deplete ozone compared with the ability of a similar amount of CFC-11.

*Source:* Gerald Mutisya, UNEP Ozone Secretariat.

# Economic Trends

© DIGITAL VISION

As it has every year for half a century, gross world product (GWP)—the tally of government estimates of the output of goods and services in nations around the world—reached a new high in 2001, at $45.9 trillion (in 2000 dollars).[1] (See Figure 1.) But the rate of growth, 2.1 percent, was among the lowest since 1950.[2] (See Figure 2.)

The current growth slowdown is the most widespread since the early 1980s, according to the International Monetary Fund.[3] It was most marked in the Americas, with growth dropping from 4.1 percent in 2000 to 1.0 percent in 2001 in the United States and nearly the same in Latin America.[4] For Western Europe, growth fell from 3.4 percent to 1.7 percent.[5] For Asia (excluding republics of the former Soviet Union), the rate fell from 5.3 percent to 2.8 percent, as a true recession—"growth" of –0.4 percent—occurred in Japan.[6]

Growth generally slowed less in poorer countries. In China, the second-largest economy, it slipped from 6.1 percent to 5.6 percent.[7] In India, the fourth-largest, it went from 6.0 percent to 4.4 percent.[8] The deceleration was also less pronounced in Eastern Europe, where the rate declined from 3.9 percent to 2.7 percent, and in the former Soviet Union, where it went from 7.5 percent (the highest rate since 1973) to 6.0 percent.[9] Growth accelerated slightly in Africa, from 3.2 percent to 3.6 percent.[10]

Overall, though, the simultaneity of the slowdown illustrates the interconnectedness of the global economy. One cause of the slowdown appears to be the rally in the world's oil markets during 2000, which sent crude prices above $30 per barrel late that year.[11]

Another is the bursting of the great technology stock bubble.[12] In the United States, the stock market peaked in total value on March 24, 2000, at about $14.5 trillion, but then lost nearly 30 percent in 12 months.[13] This, too, was part of a global phenomenon: London's FTSE 100 index fell 20 percent and Tokyo's Nikkei 225 plunged 34 percent.[14] In retrospect, perhaps trillions of dollars invested in high-technology equipment and companies went to waste. Had it been invested differently, the global economy might have grown faster in 2001.

A final common cause is the terrorist attacks of September 11, 2001, and the war in Afghanistan—but preliminary analysis suggests that these effects were and will remain relatively minor at the global level.[15] The U.S. economy had already slipped into recession six months before the attacks.[16] The 1995 earthquake in Kobe, Japan, killed more people than the 2001 terrorist attacks and did more property damage, but it had little long-term economic impact on Japan, let alone the world.[17] The attacks could deal a lasting blow to global airline and hotel industries, however, and permanently raise the cost of international commerce.

The growth rate of 2.1 percent seen in 2001 is rapid enough to double economic output every 30 years. Yet many economists consider the world economy to be in recession when it grows less than 2.5 percent a year.[18]

The global recession is bad news if it significantly slows economic development in poor countries. While income per person has climbed steadily in the industrial west since World War II, reaching an average $29,000 in 2001, it has stayed far lower in many other countries.[19] (See Figure 3.) Yet cash income is one important source of economic well-being. In poor countries, economic growth that is steady and shared by the broad mass of people is essential to development. In Africa, average gross domestic product (GDP) per person has fluctuated around $1,700 since 1973 (in 2000 dollars).[20] A demonstration of the link between GDP and poverty came in 1998, when Indonesia's economy shrank 13.7 percent and its poverty rate reportedly climbed from 11 to 18 percent.[21]

In rich countries, too, because of the way their economies work, the burden of recessions can fall on a small minority of people, often those least able to absorb the shock. Companies are much more likely to cut costs by laying off, say, 5 percent of their workers than by cutting everyone's salary 5 percent. In the United States, 1.8 million jobs disappeared in 2001 even while those looking for work expanded by 800,000—adding 2.6 million people to the ranks of the unemployed.[22]

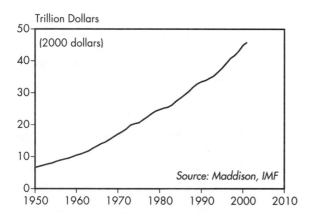

Figure 1: Gross World Product, 1950–2001

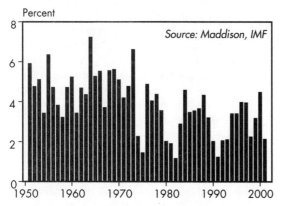

Figure 2: Growth of Gross World Product, 1951–2001

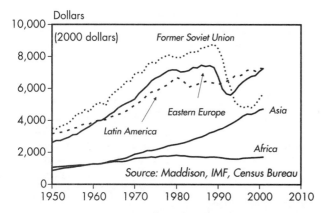

Figure 3: Gross Regional Product Per Capita, Selected Regions, 1950–2001

## Gross World Product, 1950–2001

| Year | Total (trill. 2000 dollars) | Per Person (2000 dollars) |
|---|---|---|
| 1950 | 6.6 | 2,582 |
| 1955 | 8.5 | 3,042 |
| 1960 | 10.4 | 3,438 |
| 1965 | 13.3 | 3,980 |
| 1970 | 17.1 | 4,603 |
| 1971 | 17.8 | 4,697 |
| 1972 | 18.6 | 4,823 |
| 1973 | 19.9 | 5,041 |
| 1974 | 20.3 | 5,058 |
| 1975 | 20.6 | 5,038 |
| 1976 | 21.6 | 5,191 |
| 1977 | 22.5 | 5,308 |
| 1978 | 23.4 | 5,446 |
| 1979 | 24.3 | 5,542 |
| 1980 | 24.8 | 5,557 |
| 1981 | 25.2 | 5,567 |
| 1982 | 25.5 | 5,534 |
| 1983 | 26.3 | 5,595 |
| 1984 | 27.5 | 5,753 |
| 1985 | 28.4 | 5,853 |
| 1986 | 29.4 | 5,958 |
| 1987 | 30.5 | 6,069 |
| 1988 | 31.8 | 6,224 |
| 1989 | 32.8 | 6,316 |
| 1990 | 33.5 | 6,336 |
| 1991 | 33.9 | 6,314 |
| 1992 | 34.6 | 6,346 |
| 1993 | 35.3 | 6,384 |
| 1994 | 36.5 | 6,506 |
| 1995 | 37.7 | 6,633 |
| 1996 | 39.2 | 6,803 |
| 1997 | 40.8 | 6,976 |
| 1998 | 41.7 | 7,038 |
| 1999 | 43.0 | 7,167 |
| 2000 | 44.9 | 7,392 |
| 2001 (prel) | 45.9 | 7,454 |

Sources: Worldwatch update of Angus Maddison, The World Economy: A Millennial Perspective (Paris: OECD, 2001); updates from IMF, World Economic Outlook tables.

# Trade Slows

*David Malin Roodman*

According to a preliminary estimate, the total value of world exports declined 4.1 percent in 2001—from $7.75 trillion the year before to $7.43 trillion (in 2000 dollars).[1] (See Figure 1.) This percentage drop is the largest since 1983.[2]

This drop in fact may be an underestimate, because it is based on incomplete data for late 2001, when ripple effects from the global economic slowdown and terrorist attacks began to spread. The fall in demand for jet fuel late in the year, for example, pushed down both the volume of oil exports, measured in barrels, and the price paid for each barrel, which doubly depressed the total value of oil exports.[3]

From an economic point of view, international trade occurs whenever a resident of one country sells something to a resident of another. The "something" can be a tangible good such as a barrel of oil or a car. It can also be an intangible service. When a Japanese hotel sells the use of a room for a night to a German tourist, that service counts as an export from Japan to Germany (even though the tourist traveled from Germany to Japan).

Links: pp. 30, 58, 126

International trade in goods has accelerated radically since 1950. Goods exported in 1950 were worth $380 billion (in 2000 dollars).[4] Fifty-one years later, that figure was reached every three weeks, and it totaled some $5.96 trillion for 2001 as a whole.[5]

Since 1970, exports growth for services has paralleled that for goods. From $310 billion in 1970, service exports climbed to $1.47 trillion in 2001 (in 2000 dollars).[6] Major categories of exported services in 1999 included freight (earning $134 billion), passenger transport ($83 billion), and other travel-related services ($437 billion).[7]

The ratio between the value of world trade and the value of total economic production (gross world product, or GWP) is one indicator of "globalization." Since World War II, this ratio has climbed overall. But since 1995 it actually has fallen, from a peak of 18.4 percent to 15.9 percent.[8] (See Figure 2.) Between 1996 and 1998, prices for traded goods fell 12 percent on average—rather than rising with general inflation in the U.S. dollar—mainly because of the currency crises in East Asia.[9] Prices then recovered, but the global economic slowdown took hold and reduced the physical volume of goods exports.[10]

In November 2001, diplomats met in Doha, Qatar, to launch a new round of negotiations to reduce restrictions on world trade. The previous round had lasted six years and concluded in 1994 with the creation of the World Trade Organization (WTO). If the new round indeed gets off the ground, it will be the ninth since World War II, and the most controversial yet.[11]

Especially in rich countries, many people are concerned about the way the WTO system tends to put the cause of trade liberalization ahead of important concerns such as ecological stability, protection of workplace safety standards, and human rights. In 1991, for example, the WTO ruled against a U.S. import ban on tuna caught with dolphin-ensnaring nets.[12] The law may have been good for the environment, but it was deemed harmful to trade.

At least as potent at Doha was skepticism from developing countries.[13] Historically, the United States, Western Europe, and Japan have muscled through rules that benefit their own companies more than those of poorer countries.[14] Notably, at Doha, developing countries united more than they had before, enough to extract major rhetorical concessions from richer countries.

Industrial countries promised to phase out their subsidies for agricultural exports, which glut global food markets, lower prices, and harm farmers in poor countries.[15] Delegates also endorsed a declaration stating that public health emergencies can take precedence over protecting the intellectual property of pharmaceutical companies.[16] That may make it easier for developing-country governments to break drug patents in order to obtain cheaper, copycat drugs to fight malaria, tuberculosis, and AIDS. But it will be years before negotiators hammer out what all these concessions mean in practice.

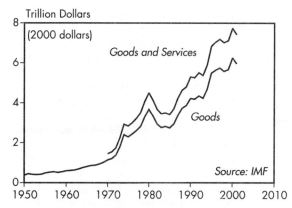

Figure 1: World Exports of Goods 1950–2001,
and Goods and Services, 1970–2001

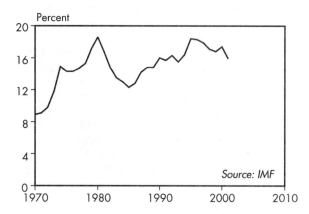

Figure 2: World Exports of Goods and Services as a
Share of Gross World Product, 1970–2001

## World Exports of Goods, 1950–2001, and Goods and Services, 1970–2001

| Year | Goods | Goods and Services |
|------|-------|--------------------|
| | | (trill. 2000 dollars) |
| 1950 | 0.38 | |
| 1955 | 0.49 | |
| 1960 | 0.60 | |
| 1965 | 0.81 | |
| 1970 | 1.14 | 1.45 |
| 1971 | 1.21 | 1.54 |
| 1972 | 1.38 | 1.74 |
| 1973 | 1.81 | 2.26 |
| 1974 | 2.40 | 2.93 |
| 1975 | 2.29 | 2.84 |
| 1976 | 2.44 | 3.00 |
| 1977 | 2.61 | 3.22 |
| 1978 | 2.82 | 3.51 |
| 1979 | 3.30 | 4.07 |
| 1980 | 3.67 | 4.50 |
| 1981 | 3.34 | 4.13 |
| 1982 | 2.93 | 3.68 |
| 1983 | 2.76 | 3.45 |
| 1984 | 2.82 | 3.49 |
| 1985 | 2.74 | 3.42 |
| 1986 | 2.94 | 3.70 |
| 1987 | 3.38 | 4.25 |
| 1988 | 3.74 | 4.64 |
| 1989 | 3.87 | 4.80 |
| 1990 | 4.22 | 5.23 |
| 1991 | 4.19 | 5.26 |
| 1992 | 4.35 | 5.51 |
| 1993 | 4.24 | 5.38 |
| 1994 | 4.69 | 5.90 |
| 1995 | 5.49 | 6.83 |
| 1996 | 5.64 | 7.05 |
| 1997 | 5.75 | 7.13 |
| 1998 | 5.58 | 7.02 |
| 1999 | 5.67 | 7.12 |
| 2000 | 6.25 | 7.75 |
| 2001 (prel) | 5.96 | 7.43 |

Source: IMF, *International Financial Statistics*, electronic database, November 2001; IMF, *World Economic Outlook Database*, December 2001.

In 2000, the cumulative foreign debt of developing and former Eastern bloc nations posted its largest one-year drop in dollar terms since detailed recordkeeping began in 1970.[1] The fall from $2.62 trillion to $2.53 trillion (in 2000 dollars) followed a smaller decline in 1999.[2] (See Figure 1.)

The drops over these two years may be statistical aberrations, however. At least 36 percent of the debt is owed in currencies other than dollars.[3] As many of those currencies fell against the dollar in 1999 and 2000, loans denominated in them shrank in the dollar-based statistics.[4] As a result, even as total debt fell in dollar terms, it rose when expressed in euros, the second-most-used currency for loans to these countries—from 2.27 trillion to 2.69 trillion (in 2000 euros).[5] Overall, the debt total is best seen not as having fallen, but as having reached a standstill after a long climb.

Link: p.118

Since loans are investments, whether high debt is good or bad for a country hinges on how well the money is used. Ideally, it supports projects—from public railroad construction to education—that ultimately boost economic output, and exports earn enough to repay the loans. In South Korea, for example, foreign lending has helped finance rapid economic development and poverty reduction.[6]

Worldwide, however, foreign funds have often been used poorly—supporting arms purchases, corruption, capital flight, and prestige projects (such as unneeded airports), in addition to more well-intended but poorly implemented projects.[7] This is one reason that countries have frequently fallen into debt trouble in recent decades, becoming unable to meet their repayment obligations. Herd mentality of investors is another.

Developing and former Eastern bloc countries divide roughly into two groups, based on the kind of debt trouble they are prone to. Middle-income countries are industrialized enough to attract serious interest from commercial creditors—bond investors and banks—and consequently borrow most heavily from them.[8] These countries accounted for 78 percent of the outstanding debt of developing and former Eastern bloc nations at the end of 1999.[9]

Middle-income countries have been struck by major debt crises at remarkably regular intervals of about 50 years since the 1820s.[10] (See Figure 2.) The most recent one hit in 1982, and sent many nations, including most of South America, into recession for nearly a decade. In Mexico, wages fell by half between 1982 and 1988.[11] In the Philippines, a million or more desperate peasants moved into the hills, where they cleared erodible slopes of protective trees and started farming to survive.[12] The last decade has seen crises in Argentina, Brazil, Ecuador, East Asia, Mexico, and Turkey.

The other debtor group consists of low-income countries such as Nicaragua and Tanzania. Generally shunned by commercial lenders, they borrow mainly from rich-world governments and other official institutions such as the World Bank and the International Monetary Fund.[13] Official creditors barely existed before World War II. Much more than commercial lenders, they are generally willing to keep lending to countries in debt trouble even if most new loans just go toward repaying old ones. Partly as a result, a historically novel form of debt trouble began to afflict many low-income countries by the 1980s—not so much crises as chronic syndromes in which new loans went largely to repaying old.[14]

Rich-world governments have enacted a series of programs since the late 1980s to reduce the debt burden on low-income countries, but the inadequacy of each has been implicitly acknowledged by the launch of the next.[15] At the end of 1999, 47 countries—37 in Africa—met the World Bank's statistical criteria for being low in income and high in debt.[16] Of these, 42 are eligible for the latest program, the Heavily Indebted Poor Countries (HIPC) initiative, which offers by far the most debt relief to date—as much as 55 percent on average for eligible countries.[17]

But even after the current HIPC program, many poor countries will probably owe more than they can pay.[18] Thus creditors are likely to bring forth yet another program to address the debt troubles of the poorest nations.

Trillion Dollars

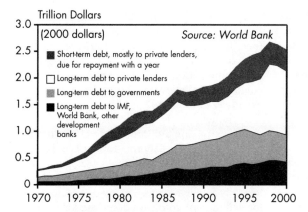

Figure 1: Foreign Debt of Developing and
Former Eastern Bloc Nations, 1970–2000

Number

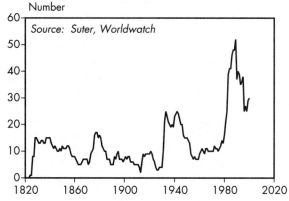

Figure 2: Number of Countries Not Servicing All
Their Foreign Debts, 1820–2000

## Foreign Debt of Developing and Former Eastern Bloc Nations, 1970–2000

| Year | Foreign Debt (trill. 2000 dollars) |
|------|------------------------------------|
| 1970 | 0.27 |
| 1971 | 0.29 |
| 1972 | 0.33 |
| 1973 | 0.38 |
| 1974 | 0.43 |
| 1975 | 0.52 |
| 1976 | 0.60 |
| 1977 | 0.75 |
| 1978 | 0.88 |
| 1979 | 1.00 |
| 1980 | 1.10 |
| 1981 | 1.20 |
| 1982 | 1.31 |
| 1983 | 1.36 |
| 1984 | 1.38 |
| 1985 | 1.50 |
| 1986 | 1.61 |
| 1987 | 1.77 |
| 1988 | 1.72 |
| 1989 | 1.74 |
| 1990 | 1.81 |
| 1991 | 1.85 |
| 1992 | 1.89 |
| 1993 | 2.02 |
| 1994 | 2.19 |
| 1995 | 2.35 |
| 1996 | 2.40 |
| 1997 | 2.45 |
| 1998 | 2.66 |
| 1999 | 2.62 |
| 2000 | 2.53 |

Source: World Bank, *Global Development Finance*, electronic database, 2001.

*Payal Sampat*

In 2001, mining companies spent just under $2 billion exploring for untapped lodes of metal around the world.[1] (See Figure 1.) This is less than half the amount spent in 1997—a record $4.2 billion.[2] Bruised by the lingering effects of the 1998 Asian financial crisis, low metals prices, and capital shortages, most mining companies have shrunk their exploration budgets and cut operating costs at existing mines.[3]

Gold has traditionally driven exploration budgets: in 1997, two thirds of all exploration was for this yellow metal.[4] But gold prices have dropped by half since 1990 (in 2000 dollars), reaching a 29-year low in 2001 and driving the share of exploration for gold down to 42 percent.[5] Metals such as copper, zinc, and nickel accounted for a greater share of exploration budgets in 2001 than they have in the past, almost 39 percent.[6]

*Links: pp. 66, 112*

As the quest for new veins of metal accelerated in the mid-1990s, most new exploration took place in the developing world. Between 1991 and 1997, exploration spending expanded six times in Latin America and almost quadrupled in the Pacific region.[7] Although spending on exploration has plummeted dramatically in all regions since then, the developing world still attracts about half of all new money.[8] Latin America remains a leading attraction, drawing 28 percent of investment in 2001.[9] (See Figure 2.) Chile, Peru, Brazil, and Mexico—all of which have courted foreign investors in the last 10 years—lead the list for this region.[10]

Southeast Asia and the Pacific has seen a 72-percent decline in investment, although multinational firms continue to operate and expand existing mines in Indonesia, Papua New Guinea, and other island nations.[11] Mining companies are keen to expand their presence in Africa—which claimed only 14 percent of all spending in 2001—in the quest for diamond and platinum deposits.[12]

The more wealthy mining regions—where most mining companies are headquartered—still maintain a strong foothold. Australia and Canada, the nations that attracted the most investment in 2001, each accounted for 17 percent of exploration spending.[13] The U.S. share, however, shrunk to just 8 percent—exploration there fell 60 percent between 1997 and 2001.[14] This cutback came in response to changes in U.S. mining laws in 2000, but many of the new environmentally favorable rules have since been revised or rolled back.[15]

Although metals serve many useful purposes, the extraction and processing of virgin minerals can impose a sobering toll on people and ecosystems. Most new mining development is taking place in some of the world's most ecologically fragile regions—many of which are located in poor countries desperate for foreign investment. These include a titanium mine in a Madagascar forest that is inhabited by rare lemurs, birds, and 20 indigenous plant species; gold exploration in Peru's Andean cloud forests; and tantalite mining in the Okapi Reserve in the Democratic Republic of Congo, home to the endangered mountain gorilla.[16]

Several studies point out that mining-dependent nations typically have sluggish rates of economic development and some of the highest poverty rates, spurring a debate about whether mining benefits poor people and countries over the long term.[17] One thing seems clear: the poor tend to bear the costs of mining disproportionately. Perhaps as much as 50 percent of gold produced between 1995 and 2015 has or will come from indigenous peoples' lands in places as diverse as Nevada and Papua (formerly Irian Jaya).[18] In Peru, local farmers have protested being displaced by the Tambo Grande mines; communities in Guyana, Papua New Guinea, and Kyrgyzstan, among others, have suffered as mines there have severely contaminated soil and water supplies.[19]

The International Labour Organization calls mining one of the most hazardous occupations. It employs just 1 percent of the global work force but is responsible for 5 percent of all worker deaths on the job—about 40 deaths a day.[20] As mining companies try to reduce operating costs, jobs in mining are in further decline. In 1999 alone, South African mines laid off about 100,000 workers—a third of the total—as operations were mechanized or closed.[21]

Billion Dollars

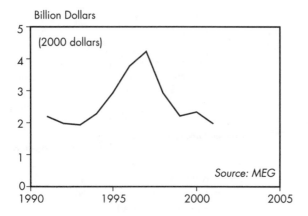

Figure 1: World Metals Exploration Investment, 1991–2001

## World Metals Exploration Investment, 1991–2001

| Year | Investment (million 2000 dollars) |
|------|-----------|
| 1991 | 2,203 |
| 1992 | 1,980 |
| 1993 | 1,934 |
| 1994 | 2,284 |
| 1995 | 2,933 |
| 1996 | 3,771 |
| 1997 | 4,230 |
| 1998 | 2,933 |
| 1999 | 2,212 |
| 2000 | 2,338 |
| 2001 | 1,966 |

Source: MEG, *Strategic Reports*, 1991–93, and press releases, 1994–2001.

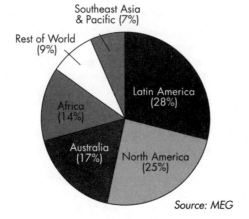

Figure 2: Share of World Metals Exploration Investment by Region, 2001

*Payal Sampat*

More than 900 million tons of metals were extracted from the earth in 2000—about 7 percent more than the previous year.[1] (See Figure 1.) In the last 30 years, a total of 24 billion tons of metals have been mined.[2] If this material were loaded onto the largest, 218-ton dump trucks that are used on mine sites, the convoy of trucks lined bumper-to-bumper could circle the globe at the Equator 34 times.[3]

Where does this enormous amount of material go once it is removed from the ground? Most of it works its way into our daily lives, forming buildings, bridges, cars, airplanes, stereos, cell phones, and other goods. Some materials, such as steel in buildings, remain in use for decades; others, such as aluminum cans, may be discarded minutes after use.

*Links: pp. 60, 64, 84, 112*

About 70 metals are mined for commercial use, including aluminum, cadmium, copper, lead, nickel, raw steel, and zinc. By weight, steel accounts for the bulk of the total—nearly 94 percent.[4]

On average, 148 kilograms of metal were produced per person in 2000.[5] (See Figure 2.) This is significantly smaller than the all-time high of 185 kilograms per person in 1973.[6] The decline reflects the expansion in global population in the last three decades, much of which has taken place in poorer regions, where materials consumption per person is relatively low.

The metals intensity of the global economy—the amount of metals used to generate economic wealth—has declined 45 percent in the last 30 years.[7] (See Figure 3.) This reflects a shift in the global economy as manufacturing and other industries that typically use large amounts of metal have grown at a far slower pace than service industries such as telecommunications and finance.

Mineral ores are unevenly distributed in Earth's crust, with some concentrated in a few regions. One third of the world's copper is extracted in Chile, for instance, while 28 percent of lead comes from China.[8] Metals are often produced in countries that are major consumers as well.[9] China, for example, is the world's largest producer and consumer of steel, while the United States produces and uses more aluminum than any other country.[10] Elsewhere, metals are extracted almost entirely for export—with even the ores sent overseas for processing and refining. For instance, Papua New Guinea and Botswana mine copper ores, but most of the output is exported to non-copper-producing countries such as South Korea and Germany to be refined.[11]

The major industrialized regions—the United States, Canada, Australia, Japan, and Western Europe—with 15 percent of the world's population, together consume 61 percent of all aluminum, 60 percent of lead, 59 percent of copper, and 49 percent of steel.[12] On a per capita basis, the different levels of consumption are especially marked: the average American uses 22 kilograms of aluminum a year, while the average for India is 2 kilograms, and for Africa, just 0.7 kilograms.[13]

For countries that are major importers or exporters of finished goods, the per capita figure may mask or overstate domestic metals use. For example, Taiwan and South Korea rank much higher than any industrial country in their copper consumption, at 29 kilograms and 18 kilograms per person.[14] But most of this feeds their large export markets for electronics and other goods.

Just a few sectors of the economy dominate metals use. In industrial countries, the transportation sector (including vehicle fleets) uses an estimated 70 percent of lead produced each year, 37 percent of steel, 33 percent of aluminum, and 27 percent of copper.[15] Construction is another major player, using 34 percent of steel, 30 percent of copper, 17 percent of lead, and 19 percent of aluminum in industrial nations.[16]

It takes far less energy to mine discarded materials than to extract, process, and refine metals from ore. It takes 95 percent less energy to produce aluminum from recycled materials, for example, than from bauxite ore.[17] Recycling copper takes seven times less energy than processing ore; recycled steel uses three-and-a-half times less.[18] Globally, 29 percent of aluminum and 13 percent of copper come from recycled sources.[19]

Million Tons

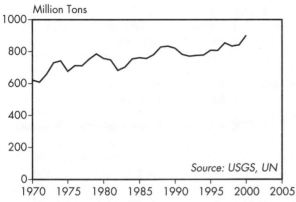

Figure 1: World Metals Production, 1970–2000

Kilograms

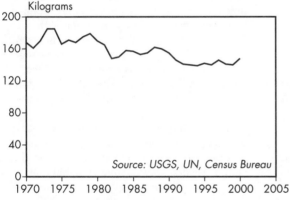

Figure 2: World Metals Use Per Person, 1970–2000

Tons Per Million Dollars

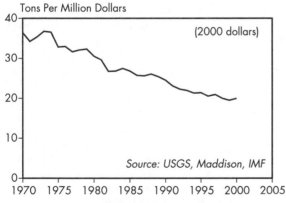

Figure 3: Metals Intensity of Global Economy, 1970–2000

## World Metals Production, 1970–2000

| Year | Metals Mined (million tons) |
| --- | --- |
| 1970 | 621 |
| 1971 | 609 |
| 1972 | 658 |
| 1973 | 730 |
| 1974 | 742 |
| 1975 | 677 |
| 1976 | 712 |
| 1977 | 711 |
| 1978 | 753 |
| 1979 | 785 |
| 1980 | 757 |
| 1981 | 748 |
| 1982 | 682 |
| 1983 | 703 |
| 1984 | 754 |
| 1985 | 761 |
| 1986 | 756 |
| 1987 | 780 |
| 1988 | 828 |
| 1989 | 833 |
| 1990 | 820 |
| 1991 | 782 |
| 1992 | 771 |
| 1993 | 775 |
| 1994 | 778 |
| 1995 | 808 |
| 1996 | 806 |
| 1997 | 854 |
| 1998 | 834 |
| 1999 | 841 |
| 2000 (prel) | 902 |

Sources: USGS, *Minerals Yearbook* and *Mineral Commodity Summaries*, various years; United Nations, *Industrial Commodities Statistics Yearbook*, various years.

The amount of oil spilled accidentally in 2000 from tankers, pipelines, wells, storage facilities, and other sources was estimated at 48,600 tons worldwide by the *Oil Spill Intelligence Report* (OSIR).[1] This was the lowest recorded since 1968. The largest amount, some 1.5 million tons, was spilled in 1979.[2] Since 1990, there has been an almost continuous reduction in the quantity of oil spilled.[3] (See Figure 1.)

These figures do not include spills that are the result of warfare or sabotage, however. Historically, three of the top five spill incidents are the result of acts of war.[4] (See Figure 2.) Attacks on oil fields and tankers during the Iran-Iraq war raised the total for 1983 by 46 percent.[5] In 1991, Iraqi troops deliberately released some 840,000 tons of oil from Kuwaiti facilities into the Persian Gulf, causing the largest marine oil spill in history.[6] And in 2000, reports of sabotage by Chechen rebels indicated that 2 million tons of oil had leaked from wells and refineries near Grozny.[7] If confirmed, this would be the largest spill ever.

*Links: pp. 38, 94*

From 1968 to 2000, there were more than 7,600 civilian incidents with about 10.6 million tons of oil spilled.[8] More than 400 war-related incidents added at least another 3.6 million tons.[9] The top 50 oil spills—just 6 percent of all incidents—account for more than half the total spillage since 1968.[10]

Oil tankers, the leading source of spills, transport some 107 million tons of oil on an average day.[11] OSIR and the International Tanker Owners Pollution Federation provide somewhat conflicting spill data for certain years.[12] (See Figure 3.) In 1968–2000, tankers, barges, and other vessels accounted for about half the total amount of oil spilled.[13] But greater use of double-hulled tankers and other safety measures have significantly reduced both the number of tanker accidents and the quantity of oil spilled.[14]

Collisions and groundings are relatively rare, but can result in large, sometimes massive, spills. The two largest tanker accidents happened off the coast of South Africa, when the *Castillo de Bellver* lost 267,000 tons in 1983, and off Brittany, France, when the *Amoco Cadiz* disgorged 234,000 tons in 1978.[15] The infamous 1989 *Exxon Valdez* incident in Alaska ranks only as the forty-second worst tanker accident in terms of quantity of oil released, although it occurred in a particularly pristine and ecologically vulnerable location.[16]

Almost half of all pipeline spills are the result of aging equipment. Some pipelines are 30–50 years old; others are even older.[17] Niger delta communities in Nigeria have suffered heavily from spills caused by corrosion of antiquated pipelines and by vandalism. Pipeline bursts have killed hundreds of people in recent years.[18]

Sabotage is another cause of pipeline spills. In the last few years rebel groups have attacked pipelines in Algeria, Assam (India), Colombia, Ecuador, Sudan, Turkey, and Yemen.[19] In Colombia, rebel groups bombed pipelines 98 times during 2000, up from 79 times during 1999.[20] Unconfirmed estimates suggest that about 43,000 tons of oil were spilled there in 2000—twice the amount lost due to all non-war pipeline incidents that year.[21]

Some well blowouts are among the biggest spills ever. From June 1979 to February 1980, for example, the Ixtoc exploratory well in the Gulf of Mexico spewed some 476,000 tons of oil, the largest non-war oil spill ever.[22] A production well in Uzbekistan's Fergana Valley spilled 299,000 tons in 1990, and one in Libya lost 143,000 tons in 1980.[23]

The quantity of oil spilled does not necessarily indicate the severity of the impact on the environment. Important factors include the type of oil spilled, weather and climate conditions, the extent to which the oil is recovered or at least contained, how quickly the oil biodegrades and how much of it evaporates, and the proximity to wildlife habitats or environmentally sensitive areas.[24]

Even though much of the oil released by the *Exxon Valdez* in 1989 evaporated or dispersed, for instance, the accident had disastrous results.[25] It killed an estimated 3,500–5,500 sea otters (10–15 percent of the region's total population) and some 300,000–675,000 seabirds.[26] Most wildlife species still have not recovered.

Thousand Tons

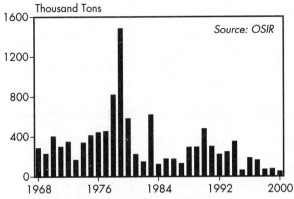

Figure 1: Oil Spills from Civilian Operations,
1968–2000

Thousand Tons

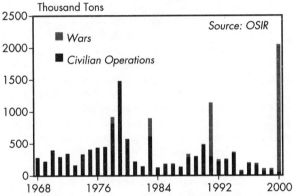

Figure 2: Oil Spills from Civilian Operations and
Wars Combined, 1968–2000

Thousand Tons

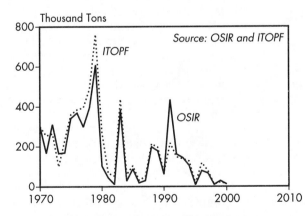

Figure 3: Tanker Oil Spills, 1970–2000

| Oil Spills from Civilian Operations, 1968–2000 | |
| --- | --- |
| Year | Oil Spilled |
| | (thousand tons) |
| 1968 | 283.4 |
| 1969 | 226.3 |
| 1970 | 399.9 |
| 1971 | 295.9 |
| 1972 | 346.2 |
| 1973 | 164.8 |
| 1974 | 336.3 |
| 1975 | 410.3 |
| 1976 | 439.6 |
| 1977 | 450.7 |
| 1978 | 816.8 |
| 1979 | 1,481.3 |
| 1980 | 577.9 |
| 1981 | 220.8 |
| 1982 | 146.3 |
| 1983 | 614.6 |
| 1984 | 120.8 |
| 1985 | 174.2 |
| 1986 | 173.3 |
| 1987 | 128.3 |
| 1988 | 290.2 |
| 1989 | 291.9 |
| 1990 | 474.4 |
| 1991 | 298.5 |
| 1992 | 218.9 |
| 1993 | 244.0 |
| 1994 | 347.7 |
| 1995 | 60.2 |
| 1996 | 183.3 |
| 1997 | 160.6 |
| 1998 | 68.2 |
| 1999 | 74.9 |
| 2000 | 48.6 |

Source: Oil Spill Intelligence Report.

# Roundwood Production Rebounds

*Janet N. Abramovitz*

According to the U.N. Food and Agriculture Organization (FAO), global production of roundwood—the logs that become fuel, lumber, paper, and other wood products—reached a new peak of 3,376 million cubic meters in 1999, the last year for which data are available.[1] (See Figure 1.) Production has topped 3,000 million cubic meters every year since 1983, more than twice the figure in 1950.[2] In the mid-1990s the global total dipped as production in the former Soviet Union fell by about two thirds during the new countries' economic transition.[3]

In 1999, 61 percent of the world's recorded wood harvest came from developing nations.[4] The share produced in industrial nations has declined from 57 percent in 1961 to 39 percent in 1999.[5] (See Figure 2.)

About 55 percent of the roundwood cut today is used directly for fuelwood and charcoal.[6] The other 45 percent becomes "industrial roundwood"—the logs that are cut into lumber and panels for construction purposes or ground into pulp to make paper.[7] Developing countries produce about 89 percent of wood cut specifically for fuel.[8] But these figures are misleading in terms of the importance of wood fuel in industrial countries: where there are large forest products industries, by-products such as wood chips and sawdust are burned to fuel the mills. These add close to 300 million cubic meters of wood to the 173 million used directly for fuel in industrial countries.[9]

Link: p. 104

The industrial roundwood harvest has remained concentrated in just five countries since the 1970s: the United States, Canada, Russia, China, and Brazil. These five produce 58 percent of the world's recorded production. Together the top 10 (adding in Sweden, Finland, Germany, France, and Indonesia) accounted for about 72 percent of production.[10]

Industrial nations produce 73 percent of industrial roundwood, a share that has declined since 1970 as developing nations expanded their output.[11] While production in industrial nations has remained relatively constant since 1970, in developing nations it has doubled.[12] Industrial nations continue to consume a disproportionate share of global production—77 percent of the timber harvested for industrial use is consumed by the 22 percent of the world living in industrial nations.[13] Although the United States uses the most, China is now second.[14]

Production of some industrial wood products has grown more rapidly than others. Between 1961 and 1999, paper production grew by 309 percent.[15] Paper and paperboard now account for the largest single share of industrial wood use, at 40 percent, through wood cut directly for paper and the use of residues from other wood processing mills.[16] Sawnwood, the lumber used for construction and furniture, dropped from 34 percent of production in 1961 to 27 percent in 1999.[17] Total sawnwood production increased by only 18 percent since 1961, and has declined from peak production in the late 1980s.[18] (See Figure 3.) Production of wood panels like plywood (which have replaced sawnwood in some cases) jumped 545 percent since 1961, now accounting for 11 percent of production.[19]

Due to illegal production and trade, output data are reported by governments to the FAO and may not reflect full levels of production. In Indonesia, for example, an independent study by the U.K. Department for International Development found that production was more than double the amount reported by the government.[20] Extensive illegal harvest and trade have also been reported in Brazil, Russia, Cambodia, Liberia, Papua New Guinea, Cameroon, and elsewhere.[21] Growing recognition of this widespread problem is beginning to spur government commitments to combat illegal logging and trade.[22]

The area of commercial forest certified as well-managed has grown substantially in recent years. By the end of 2001, over 25 million hectares had been certified to Forest Stewardship Council (FSC) standards, more than double the area in 1998.[23] While there are FSC-certified forests in 54 countries, 67 percent of the acreage is in Europe and 13 percent is in North America.[24] Wood products originating in well-managed forests are still only a small share of the market.

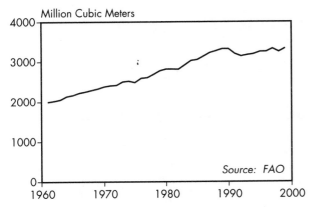

Figure 1: World Roundwood Production, 1961–99

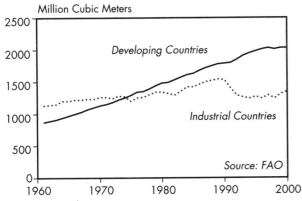

Figure 2: World Roundwood Production, Industrial
and Developing Countries, 1961–99

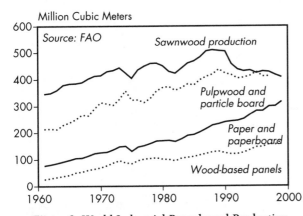

Figure 3: World Industrial Roundwood Production,
by Type, 1961–99

## World Roundwood Production, 1961–99

| Year | Production (million cubic meters) |
|------|-----------------------------------|
| 1961 | 2,000 |
| 1962 | 2,023 |
| 1963 | 2,055 |
| 1964 | 2,137 |
| 1965 | 2,168 |
| 1966 | 2,223 |
| 1967 | 2,254 |
| 1968 | 2,293 |
| 1969 | 2,330 |
| 1970 | 2,381 |
| 1971 | 2,407 |
| 1972 | 2,418 |
| 1973 | 2,502 |
| 1974 | 2,520 |
| 1975 | 2,486 |
| 1976 | 2,591 |
| 1977 | 2,608 |
| 1978 | 2,689 |
| 1979 | 2,776 |
| 1980 | 2,818 |
| 1981 | 2,817 |
| 1982 | 2,813 |
| 1983 | 2,921 |
| 1984 | 3,030 |
| 1985 | 3,053 |
| 1986 | 3,137 |
| 1987 | 3,224 |
| 1988 | 3,270 |
| 1989 | 3,321 |
| 1990 | 3,320 |
| 1991 | 3,201 |
| 1992 | 3,140 |
| 1993 | 3,170 |
| 1994 | 3,193 |
| 1995 | 3,250 |
| 1996 | 3,254 |
| 1997 | 3,328 |
| 1998 | 3,251 |
| 1999 | 3,336 |

Source: FAO, FAOSTAT Statistics Database, at <app.fao.org>, updated 7 November 2001.

# Transportation Trends

Vehicle Production Declines Slightly

Bicycle Production Rolls Forward

Passenger Rail at Crossroads

According to DRI-WEFA Global Automotive Group estimates, global passenger car production declined 2.7 percent in 2001, to 40 million units.[1] (See Figure 1.) Light truck production also declined slightly, to 15 million.[2] Global passenger car production outpaced sales by about 1.3 million vehicles, but sales of light trucks surpassed production by about 1.7 million.[3] The global passenger car fleet grew to 555 million in 2001.[4] (See Figure 2.)

The global auto industry continues to suffer from substantial overcapacity. Analysts at PricewaterhouseCoopers estimated global capacity to manufacture passenger cars and light trucks in 2001 at 77.3 million units, but only about 70 percent of capacity is in use.[5] At 78 percent, capacity utilization in North America and Western Europe is far higher than elsewhere.[6]

Links: pp. 150, 152

After shedding weight in the 1980s, cars have gotten heavier again in the 1990s, even though manufacturers made increasing use of light materials like plastic and aluminum. A typical U.S. family vehicle weighed 1,619 kilograms (kg) in 1978 and then 1,424 kg in 1990, but 1,501 kg in 2001.[7] The motor vehicle industry's appetite for materials remains considerable, although at least 75 percent of a car's material content ends up being recycled.[8] In the United States, the industry accounted for 33 percent of aluminum use in 2000, up from 17 percent in 1991.[9] In recent years, the industry has accounted for 70–80 percent of U.S. natural rubber consumption, 65–77 percent of lead, 55–64 percent of synthetic rubber, one third of iron, 23 percent of zinc, about 15 percent of steel, and 12 percent of copper.[10]

The industry also uses substantial amounts of energy, but far more is consumed in operating vehicles than in manufacturing them. Advances in fuel efficiency would have led to reduced gasoline consumption from car use had it not been for a variety of offsetting trends such as larger cars and more powerful engines, an ever expanding car fleet, and continuous growth in distances traveled.

The United States has slightly more than one quarter of the world's passenger cars.[11] The fuel economy of new cars improved from just 14.2 miles per gallon (equivalent to 16.6 liters per 100 kilometers) in 1974 to 28.8 miles per gallon in 1988.[12] But instead of additional progress, there has been some backsliding since then.[13] The combined fuel economy of new passenger cars and light trucks reached a high of 26.7 miles per gallon in 1987, but now stands at just 24.7, the second-lowest figure in 20 years.[14]

Since the mid-1980s, fuel efficiency has leveled off or declined in most other industrial countries as well.[15] But fuel economy in Europe (particularly in France and Italy) and Japan remains higher than in the United States, where the popularity of light trucks makes improved efficiency an elusive target.[16] (See Figure 3.) Because European and Japanese fuel economy tests use tougher methods, their results may actually be as much as 18 percent lower than they would be in the United States.[17]

Since the late 1990s, fresh gains in fuel economy have been achieved.[18] In Japan, regulations will likely bring about a rise to about 35 miles per gallon (6.7 liters per 100 kilometers) for new models by 2010.[19] The European Automobile Manufacturers Association has offered a voluntary commitment to reach 41 miles per gallon by 2008.[20]

Even though a recent U.S. National Academy of Sciences panel found that fuel economy could be raised 16–47 percent over the next 10–15 years, U.S. carmakers show little interest.[21] Steven Plotkin of the Argonne National Laboratory expects U.S. fuel economy to be no higher than 25.6 miles per gallon by 2010.[22] A program initiated in 1993 to develop 80-miles-per-gallon cars by 2004 fell short of expectations. It was abandoned by the Bush administration in early 2002 in favor of pursuing hydrogen-based fuel cell cars—which is unlikely to bear fruit for 10–20 years.[23]

Hybrid gas-electric vehicles occupy only a tiny market niche so far, although they get easily twice the fuel economy of a standard car. Toyota is planning to produce 300,00 hybrids a year by 2005, less than 1 percent of current car production.[24]

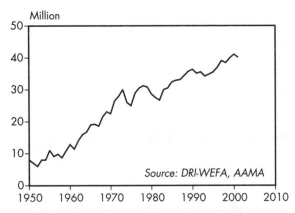

Figure 1: World Automobile Production, 1950-2001

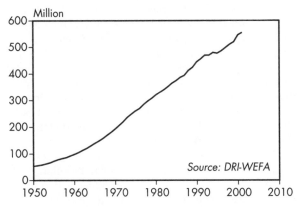

Figure 2: World Passenger Car Fleet, 1950-2001

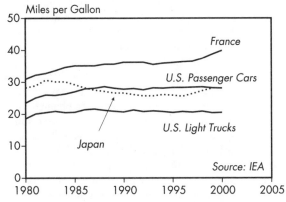

Figure 3: New Car Fuel Economy, Selected
Industrial Countries, 1980–2000

## World Automobile Production, 1950–2001

| Year | Production (million) |
|---|---|
| 1950 | 8.0 |
| 1955 | 11.0 |
| 1960 | 12.8 |
| 1965 | 19.0 |
| 1970 | 22.5 |
| 1971 | 26.5 |
| 1972 | 27.9 |
| 1973 | 30.0 |
| 1974 | 26.0 |
| 1975 | 25.0 |
| 1976 | 28.9 |
| 1977 | 30.5 |
| 1978 | 31.2 |
| 1979 | 30.8 |
| 1980 | 28.6 |
| 1981 | 27.5 |
| 1982 | 26.7 |
| 1983 | 30.0 |
| 1984 | 30.5 |
| 1985 | 32.4 |
| 1986 | 32.9 |
| 1987 | 33.1 |
| 1988 | 34.4 |
| 1989 | 35.7 |
| 1990 | 36.3 |
| 1991 | 35.1 |
| 1992 | 35.5 |
| 1993 | 34.2 |
| 1994 | 34.8 |
| 1995 | 35.5 |
| 1996 | 36.9 |
| 1997 | 39.1 |
| 1998 | 38.4 |
| 1999 | 39.9 |
| 2000 | 41.1 |
| 2001 (prel) | 40.0 |

*Sources:* DRI-WEFA Global Automotive
Group; American Automobile Manufacturers Association.

*Gary Gardner*

Production of bicycles topped 100 million units in 2000, the last year for which global data can be estimated.[1] (See Figure 1.) The nearly 9-percent increase over 1999, while robust, returns global production only to the levels of the early 1990s.[2] Globally, the industry continues to struggle and to become more concentrated.

Nearly all of the increase in 2000 came from China, where production reached 52 million units, up from 43 million in 1999.[3] For the first time, China accounts for more than half of global output.[4] (See Figure 2.) The other major Asian players—India, Taiwan, and Japan—saw production stagnate or decline.[5] Meanwhile, the European Union, the other major production center, saw output increase by a modest 3.6 percent.[6]

Production in the United States, once a significant source of bicycles, has slipped steadily from 8.5 million units in 1995 to 1.1 million in 2000.[7] But the country strengthened its place as the world's largest market in 2001, with purchases totaling more than 20 million units—one fifth of global production, and 15 percent more than in 2000.[8] The United States now imports more than 95 percent of the bicycles it uses.[9]

Link: p. 152

Indeed, a map of global bicycle flows would reveal bulging arrows from China to the rest of the world, especially the United States, and increasingly anemic arrows emanating from many other producers. Of the roughly 46 countries with bicycle production data for 1995, more than a third have seen steady declines in production since then, even as global production recovered.[10] Production increases have been most notable in low-wage nations such as China, Mexico, and Viet Nam.[11]

Bicycle use is influenced by government policy and changes in technology, among other factors. Municipal leadership in construction and promotion of a 300-kilometer-long network of bicycle paths in Bogota, Colombia, for example, is credited with boosting the cycling share of the city's population from 0.5 percent in 1997 to more than 5 percent today—more than five times the levels found in many car-centric countries such as the United States.[12]

Santiago, Chile, is following suit as it undertakes a 30–40 kilometer pilot project with funding from the Global Environment Facility.[13] As a way to combat the city's notorious air pollution, the project could grow over 10 years into a 1,000-kilometer network if city plans are fully implemented.[14]

Such investments can help reduce the dangers of cycling, a major impediment to bicycle use. In surveys in three U.S. cities in the early 1990s, more than half of respondents cited lack of safety as an influential factor in their decisions not to cycle.[15] Indeed, cycling fatalities per kilometer traveled in the United States are 11 times higher than fatalities from driving.[16] By contrast, cycling deaths in the Netherlands and Germany, where cycling-oriented laws and infrastructure are widespread, are about a quarter the level found in the United States.[17]

Emerging technologies could also affect cycling trends. Sales of electric bicycles have grown rapidly since their debut in the early 1990s, jumping by 27 percent in 2001 alone.[18] (See Figure 3.) Though this is less than 1 percent of global bicycle production, growth could continue to be brisk as batteries become lighter and more powerful and as the advantages of electrics become better known. By helping riders to go farther and cover hillier terrain than many would on a conventional bicycle, electrics have the potential to broaden interest in cycling. One industry consultant says it is "entirely possible" that the majority of bikes sold 10 years from now will have an electric drive of some sort.[19]

In summer 2000, a firm called Manhattan Scientics unveiled an electric bicycle powered by a fuel cell rather than a battery.[20] If successful, it could eliminate the technology's major environmental and performance blemish: dependence on toxic batteries that have short operational lives. This bike runs on hydrogen, the most abundant element in the universe, and a fuel whose only byproduct is water vapor. It is also due to weigh less and run longer before refueling than today's battery-powered electrics.[21] The company expects to have the new bike on the market in 2003.[22]

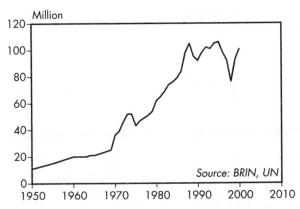

Figure 1: World Bicycle Production, 1950–2000

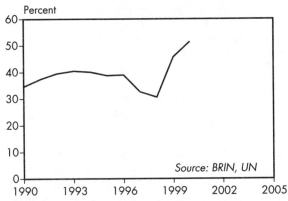

Figure 2: Chinese Bicycle Production as a Share of
World Production, 1990–2000

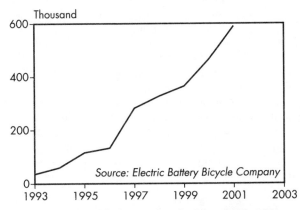

Figure 3: World Electric Bicycle Sales, 1993–2001

## World Bicycle Production, 1950–2000

| Year | Production (million) |
|---|---|
| 1950 | 11 |
| 1955 | 15 |
| 1960 | 20 |
| 1965 | 21 |
| 1970 | 36 |
| 1971 | 39 |
| 1972 | 46 |
| 1973 | 52 |
| 1974 | 52 |
| 1975 | 43 |
| 1976 | 47 |
| 1977 | 49 |
| 1978 | 51 |
| 1979 | 54 |
| 1980 | 62 |
| 1981 | 65 |
| 1982 | 69 |
| 1983 | 74 |
| 1984 | 76 |
| 1985 | 79 |
| 1986 | 84 |
| 1987 | 98 |
| 1988 | 105 |
| 1989 | 95 |
| 1990 | 92 |
| 1991 | 98 |
| 1992 | 102 |
| 1993 | 101 |
| 1994 | 105 |
| 1995 | 106 |
| 1996 | 98 |
| 1997 | 92 |
| 1998 | 76 |
| 1999 | 93 |
| 2000 (prel) | 101 |

Sources: Bicycle Retailer and Industry News,
Industry Directory 2002; United Nations,
Industrial Commodity Statistics Yearbook,
1999.

# Passenger Rail at Crossroads

*Molly O. Sheehan*

Between 1988 and 1999, world rail travel stagnated at about 1.8 trillion passenger-kilometers.[1] (See Figure 1.) As the total volume of passenger travel grew, rail's share decreased in relation to road and air.[2]

The global number masks huge national differences. More than 1 million kilometers of tracks crisscross some 120 nations, but most train travel is in the former Soviet states, India, China, Western Europe, and Japan, which together account for more than 80 percent of all passenger-kilometers.[3] (See Figure 2.) Railroads in Western Europe and Japan are geared toward passenger service, whereas extensive rail networks in the United States and Canada are used primarily for freight.[4]

Links: pp. 74, 152

The role of railroads in world transport is constantly evolving. After the first train ran in England in 1825, rail grew so rapidly that by 1900 it accounted for close to 90 percent of all passenger traffic in Europe and the United States.[5] Once cars and planes developed markets, however, trains lost passengers. Today, rail is poised for a renaissance as demand for transportation rises, particularly in developing countries, and as industrial nations seek greener alternatives to clogged airports and roads.

Planes make more sense for long distances, and cars, transit, and bicycles for shorter trips. But over 50–1,000 kilometers (30–600 miles), trains with enough passengers can be cheaper, more comfortable, and less polluting, given the high costs of flying large jets short distances and the high per capita fuel use and space required for automobiles.[6]

High-speed rail has begun to fill this niche in Japan and Western Europe.[7] Initially funded in part by World Bank loans, Japan's *shinkansen*, or "bullet train," opened in 1964 and linked Tokyo and Osaka.[8] It has since been expanded and upgraded. When France's fast train, the TGV, debuted in 1981, it cut the trip between Paris and Lyons from four to two hours; within a month, planes lost half their passengers on that route, and car traffic between those cities dropped by a third.[9] Today, passengers on a United Airlines "flight"

from Washington, DC, to Lyons connect at the Paris airport to the TGV for the final leg of their journey.[10] Germany's ICE, introduced in 1991, prompted Lufthansa to stop flying between Hannover and Frankfurt.[11] And in 2001, the new Thalys train led Air France to cancel its Paris-to-Brussels flights.[12]

Many of the world's rail passengers live in developing Asia, where rail promises to efficiently connect dense urban centers. China plans to boost its rail network and has lifted restrictions on foreign investors.[13] Between 1997 and 2000, Chinese railways raised speeds three times and started scheduling more overnight trains.[14] Future plans include a high-speed link between Shanghai and Beijing, a distance equal to the combined French and German high-speed tracks.[15] Elsewhere, South Korea is building a high-speed rail link, and Taiwan is planning one as well.[16]

While Japan's private rail network and France's public one both excel, many nations are struggling to find the best formula for them.[17] In the United States, the government subsidized Amtrak to provide national rail service in 1971, but the company has yet to develop the quality of service needed to boost revenues sufficiently.[18] After the United Kingdom divided and sold its state-run network in 1994, serious accidents showed that the new owner, Railtrack, was not maintaining the tracks well; repairs caused huge delays, prompting passengers to flee.[19] The debacle has made officials in countries such as Germany slow their privatization plans.[20] World Bank Railways Adviser Lou Thompson concludes that rail systems would work best if they were publicly defined and supported, but privately operated.[21]

Whether private or public, operators must improve service to achieve rail's people-moving potential. Train travel could be made quicker, for instance, if there were global standards for railway equipment that would ease trans-border travel, as well as advances in technology.[22]

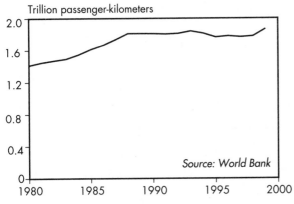

Trillion passenger-kilometers

Source: World Bank

Figure 1: World Passenger Rail Travel, 1980–99

| World Passenger Rail Travel, 1980–99 | |
| --- | --- |
| Year | Passenger-kilometers (trillion) |
| 1980 | 1.4 |
| 1981 | 1.4 |
| 1982 | 1.5 |
| 1983 | 1.5 |
| 1984 | 1.5 |
| 1985 | 1.6 |
| 1986 | 1.7 |
| 1987 | 1.7 |
| 1988 | 1.8 |
| 1989 | 1.8 |
| 1990 | 1.8 |
| 1991 | 1.8 |
| 1992 | 1.8 |
| 1993 | 1.8 |
| 1994 | 1.8 |
| 1995 | 1.8 |
| 1996 | 1.8 |
| 1997 | 1.8 |
| 1998 | 1.8 |
| 1999 | 1.9 |

Source: World Bank, "Railways Database"; Louis Thompson, Railways Advisor, World Bank.

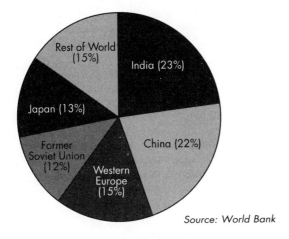

Source: World Bank

Figure 2: World's Passenger Rail Ridership, by Region or Country, 1999

# Communications Trends

Internet Continues Meteoric Rise

Mobile Phone Use Booms

In 2001, about 520 million people used the Internet, linked by a global network of 147 million host computers.[1] (See Figure 1.) The Internet has almost doubled in size since 1999, although since 1996 it has been growing more slowly than it did initially.[2] Today, 1 in every 12 people in the world goes online to get news, send e-mail, buy goods, or be entertained.[3]

The United States, where the Internet was developed, continues to dominate this electronic network. About a third of all people online are American—some 166 million.[4] (See Link: p. 112 Figure 2.) In the last two years, Japan's Internet users have doubled in size to 47 million.[5] And in China, almost 34 million people used the Internet in 2001, nearly four times more than in 1999.[6] Today, six times more Chinese use the Internet than own cars.[7] South Korea has expanded its online numbers just as rapidly, going from just 6 million in 1999 to 22 million two years later.[8]

In nine wired nations, more than half the population uses the Internet.[9] (See Figure 3.) Sweden leads this category, with 63 percent online; Iceland, Denmark, and the Netherlands are also on this list.[10] Most people in Hong Kong, Singapore, and Taiwan go online regularly.[11] In the more populated Asian countries, however, just a small share of people have access to the Internet: 2.6 percent of China, 1 percent of Indonesia, and less than 0.5 percent of India, for example.[12] More people in Singapore use the Internet than in all of Indonesia—a country with 50 times as many people.[13]

One in five Internet users lives in the developing world—about 100 million people.[14] Of the 25 million online in Latin America, nearly half live in Brazil.[15] An additional 4 million are in Argentina, and 3.4 million in Mexico.[16] But most of Africa is left out of this global network, still beleaguered by the lack of infrastructure, particularly telephone lines, and high connection costs. Even today, just 4 million Africans have Internet access—2.4 million in South Africa, and another 600,000 in Egypt—just a little more than the online population of Hong Kong.[17]

English is still the primary language used online, but for the first time ever, in 2001 the majority of people (292 million) using the Internet were non-English speakers.[18] Nearly 32 percent of them use European languages, led by German and Spanish, while 25 percent use Asian languages such as Japanese, Chinese, and Korean.[19] Forecasters estimate that by 2007 Chinese will be the most widely used language on the Internet.[20]

The value of many Internet stocks took a tumble in 2001, dampening the growth of online commercial activity. Globally, e-commerce reached $600 billion in 2001—which is 68 percent more than spent in 2000, but well below levels forecast before the economic downslide.[21] About 40 percent of this total was spent in the United States, and another 10 percent in Japan.[22] In the United States, $4 billion was spent on advertising online in 2001, accounting for some 4 percent of the nation's advertising budget.[23]

At 100 trillion bytes, the World Wide Web stores five times more data than the U.S. Library of Congress—although the quality of information is often dubious.[24] At last count, there were 10 billion pages on the Web, an 11-fold expansion since 1998.[25]

Although the Internet is making only slow inroads in some of the poorest parts of the world, it can be extremely useful when it does get there. Telemedicine projects in Mozambique, Uganda, and Bangladesh have improved medical care in remote and poorly equipped areas. Using low-cost equipment, rural doctors can send X-rays or laboratory results to medical experts at hospitals in larger cities, and get advice about treatment.[26] At 20 learning centers in India and in Morocco, primary school teachers are getting long-distance training over single terminal hookups.[27]

Unfortunately, the wired world is generating piles of hazardous electronic wastes: a computer monitor, for instance, contains four to eight pounds of lead.[28] Some 50–80 percent of used computers, circuit boards, and monitors discarded in the United States are sent to China, India, and Pakistan for recycling and disposal, exposing workers to toxins and poisoning groundwater supplies.[29]

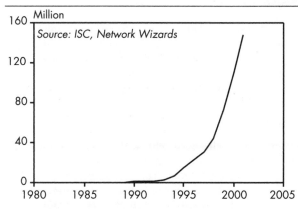

Figure 1: Internet Host Computers, 1981–2001

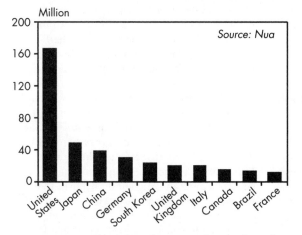

Figure 2: Top 10 Wired Nations, by Number of
Internet Users, 2001

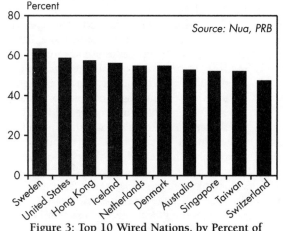

Figure 3: Top 10 Wired Nations, by Percent of
Population Online, 2001

## Internet Host Computers, 1981–2001

| Year | Host Computers |
|------|---------------|
|      | (number) |
| 1981 | 213 |
| 1982 | 235 |
| 1983 | 562 |
| 1984 | 1,024 |
| 1985 | 2,308 |
| 1986 | 5,089 |
| 1987 | 28,174 |
| 1988 | 80,000 |
| 1989 | 159,000 |
| 1990 | 376,000 |
| 1991 | 727,000 |
| 1992 | 1,313,000 |
| 1993 | 2,217,000 |
| 1994 | 5,846,000 |
| 1995 | 14,352,000 |
| 1996 | 21,819,000 |
| 1997 | 29,670,000 |
| 1998 | 43,230,000 |
| 1999 | 72,398,092 |
| 2000 | 109,574,429 |
| 2001 | 147,344,723 |

*Source:* Internet Software Consortium and Network Wizards.

# Mobile Phone Use Booms

*Molly O. Sheehan*

The number of cellular or mobile telephone subscribers rose 38 percent to nearly 1 billion in 2001, according to the International Telecommunication Union (ITU), a specialized U.N. agency charged with fostering common global telecom policies.[1] (See Figure 1.) Mobile subscribers worldwide doubled every 20 months during the 1990s.[2]

While most mobile phones are owned by people with access to conventional, fixed-line phone service, for a growing number of people in the developing world they are the sole communications tool.[3] As a result, the cellular phone boom is swelling the total number of people with access to phone service. It took 100 years to connect the first billion people by phone, but only 10 years for the second billion.[4] The ITU forecasts that at some point in 2002, the number of cellular subscribers will surpass the number of fixed-line connections, which stood at 1.045 billion in 2001.[5] (See Figure 2.)

Links: pp. 110, 112

Some 40 percent of the world's mobile phone users are in Europe, and 34 percent are in Asia.[6] The largest manufacturer of mobile phones, Nokia, is based in Finland, where cell phones dominate the economy.[7] As some markets in Western Europe reached saturation in 2001 (see Figure 3), a slowdown in demand caused global shipments of cell phones to decline.[8]

There is still considerable room for growth, however, in the world's largest markets. The United States, with more than 109 million cellular subscribers, had more mobile phones in use than any other nation in 2000 but less than 40 mobiles per 100 people.[9] Contracts that charge subscribers for incoming as well as outgoing calls may have dampened growth.[10]

China was the second largest market in 2000, with 85 million subscribers, but less than 7 mobile phones for every 100 people.[11] The number of mobile subscribers in China grew on average 85 percent a year between 1996 and 1999; China Mobile has more subscribers than any other cellular phone company in the world.[12]

In general, the greatest growth is occurring in developing countries, where prepaid phone cards have become popular for use with mobiles. These reduce the risk to the phone companies and allow people to use cellulars who do not have sufficient credit to qualify for conventional phone service.[13] In Latin America, where prepaid services prevail, the number of new mobile users has exceeded new subscribers to fixed-line services each year since 1997; one in four phone users in the region now relies on a cellular.[14]

In Africa, the number of mobile phones surpassed the number of fixed-line connections in 2001.[15] Four out of five subscribers use prepaid cards.[16] Between 1995 and 2001, the number of mobile networks in Africa grew from 33 to 100, as the number of countries without a mobile network shrunk from 28 to just 6.[17] Although in 1998 only Finland and Cambodia had more mobile subscribers than fixed lines, by the end of 2000 some 38 countries were in this category—and 20 were in Africa.[18]

Technologies and policies that promote cell phone use can benefit poor people. In 2001, a company developed a wind-up mobile phone charger that is well suited to rural areas of the developing world that lack reliable power.[19] Muhammad Yunus, the founder of the Grameen Bank in Bangladesh, believes that loans for small communications businesses can empower people.[20] Since 1997, Grameen Telecom has sold some 2,200 mobile phones to rural entrepreneurs in Bangladesh, mainly women, who in turn sell phone services to their neighbors.[21]

There are drawbacks, however, to increased reliance on mobile phones. For instance, using them while driving poses a hazard on the roads.[22] Discarded cell phones are a growing contributor to electronic waste, as consumers seek the latest technology and some manufacturers introduce disposable models.[23] Finally, researchers continue to ask whether the radio waves emitted by cell phones harm humans, particularly children whose thinner skulls and developing nervous systems make them more vulnerable.[24] In January 2002, the United Kingdom announced several research projects coordinated by the World Health Organization to further investigate this issue.[25]

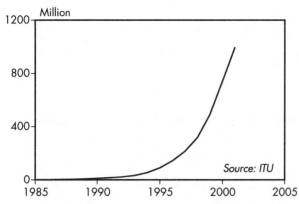

Figure 1: Cellular Telephone Subscribers
Worldwide, 1985–2001

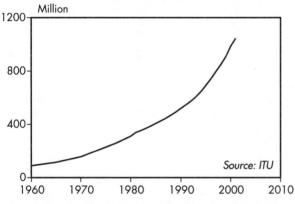

Figure 2: Telephone Lines Worldwide, 1960–2001

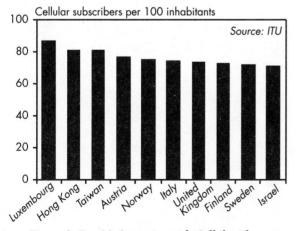

Figure 3: Top 10 Countries with Cellular Phones
Per Person, 2000

## Telephone Lines and Cellular Phone Subscribers Worldwide, 1960–2001

| Year | Telephone Lines | Cellular Phone Subscribers |
|------|------|------|
| | (million) | |
| 1960 | 89 | – |
| 1965 | 115 | – |
| 1970 | 156 | – |
| 1975 | 229 | – |
| 1976 | 244 | – |
| 1977 | 259 | – |
| 1978 | 276 | – |
| 1979 | 294 | – |
| 1980 | 311 | – |
| 1981 | 339 | – |
| 1982 | 354 | – |
| 1983 | 370 | – |
| 1984 | 388 | – |
| 1985 | 407 | 1 |
| 1986 | 426 | 1 |
| 1987 | 446 | 2 |
| 1988 | 469 | 4 |
| 1989 | 493 | 7 |
| 1990 | 519 | 11 |
| 1991 | 545 | 16 |
| 1992 | 573 | 23 |
| 1993 | 606 | 34 |
| 1994 | 646 | 56 |
| 1995 | 692 | 91 |
| 1996 | 741 | 144 |
| 1997 | 781 | 215 |
| 1998 | 849 | 319 |
| 1999 | 907 | 491 |
| 2000 | 986 | 741 |
| 2001 (prel) | 1,045 | 995 |

Source: ITU, press release, 8 February 2002; ITU, "Cellular Subscribers," 9 January 2002; ITU, STARS database.

# Health and Social Trends

HARVEY NELSON, M/MC PHOTOSHARE, WWW.JHUCCP.ORG/MMC

Population Growing Steadily

AIDS Passes 20-Year Mark

The world's population swelled to 6.2 billion in 2001—more than double the number in 1950.[1] (See Figure 1.) This represents an increase of 77 million people over the preceding year, roughly the equivalent of another Germany.[2] (See Figure 2.)

More than 95 percent of this growth is occurring in the developing world. And most of the people are added in just a handful of countries—India and China alone account for over one third of the growth.[3]

Africa has the highest growth rate of any region, increasing by 2.4 percent each year.[4] Population there is expected to more than double—from 800 million to 2.3 billion—by 2050.[5] Growth rates in Asia are lower, but they apply to a much larger base.[6] More than half of the world's people—3.7 billion—live in Asia.[7] In South Central Asia, which includes India, Pakistan, Bangladesh, and Afghanistan, population is projected to double from the current 1.5 billion by mid-century.[8]

Link: p. 148

While population in developing nations continues to rise, many industrial nations have low fertility rates. In Armenia, Italy, Spain, the Ukraine, and Russia—where the average woman bears 1.2 children in her lifetime—the low number of births has sparked concern about how these nations will adjust to aging populations and a smaller work force.[9]

The global rate of population growth has actually decreased over the past three decades—from 2.1 percent a year in 1970 to under 1.3 percent today.[10] (See Figure 3.) But this does not mean that population growth is on the decline. In fact, the number of people added to the planet each year is near the all-time high reached in the late 1980s.[11]

In the regions of the world where population continues to grow, the increase is largely caused by a combination of poverty, discrimination and violence against women, and unmet needs for reproductive health care. The United Nations reports that the annual population growth rate in "more developed" nations is just 0.3 percent, compared with 1.62 percent in "less developed" nations.[12]

And the "least developed" nations, predominantly in Africa, are growing at 2.5 percent each year.[13]

Rapid population growth makes it hard to increase living standards. Many cities in the developing world have doubled their populations in just the past 15 years, straining their capacity to provide schooling, heath care, and jobs to growing generations.[14]

Although contraceptive use has grown sixfold over the past 40 years—from just 10 percent of couples in 1960 to 60 percent in 2000—there are still barriers preventing women from planning pregnancies.[15] In some sub-Saharan African nations, birth control costs 20 percent of the average income.[16] And sexual violence often leads to unwanted pregnancy—one study in Nicaragua found that abused women are twice as likely as other women to have four or more children.[17]

An estimated 125 million women do not want to be pregnant but are not using any type of contraception.[18] Overall, 350 million women lack any access to family planning services.[19] In addition, the "global gag rule"—the U.S. administration's block on aid to international agencies that advocate or counsel patients about abortion—and a shortage of contraceptives worldwide limit the choices women and couples can make about family size.[20]

This unmet need is likely to grow, exacerbated by growth in the number of young people worldwide and a growing desire to delay childbearing. The largest generation of young people in human history—1.7 billion people aged 10–24—is now reaching reproductive age.[21] Today, 525 million women use contraception, a number projected to reach 742 million by 2015.[22]

But halting population growth is not just about controlling births. Gender inequity in education, politics, and employment prevents women from controlling their own fertility. Only 52 percent of girls in "least developed" nations stay in school after grade 4, and most of the world's illiterates are women.[23] Women are still vastly outnumbered by men at all levels of government.[24]

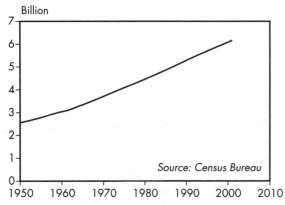

Figure 1: World Population, 1950–2001

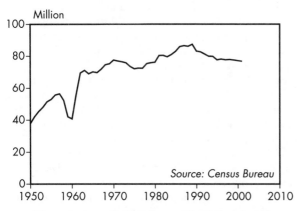

Figure 2: Annual Addition to World Population,
1950–2001

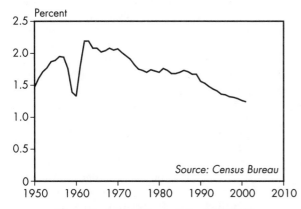

Figure 3: Annual Growth Rate of World
Population, 1950–2001

## World Population, Total and Annual Addition, 1950–2001

| Year | Total* (billion) | Annual Addition (million) |
|------|------------------|---------------------------|
| 1950 | 2.555 | 38 |
| 1955 | 2.780 | 53 |
| 1960 | 3.039 | 41 |
| 1965 | 3.346 | 70 |
| 1970 | 3.708 | 78 |
| 1971 | 3.785 | 77 |
| 1972 | 3.862 | 77 |
| 1973 | 3.939 | 76 |
| 1974 | 4.015 | 74 |
| 1975 | 4.088 | 72 |
| 1976 | 4.160 | 73 |
| 1977 | 4.233 | 72 |
| 1978 | 4.305 | 75 |
| 1979 | 4.381 | 76 |
| 1980 | 4.457 | 76 |
| 1981 | 4.533 | 80 |
| 1982 | 4.613 | 81 |
| 1983 | 4.694 | 80 |
| 1984 | 4.774 | 81 |
| 1985 | 4.855 | 83 |
| 1986 | 4.938 | 86 |
| 1987 | 5.024 | 87 |
| 1988 | 5.110 | 86 |
| 1989 | 5.196 | 87 |
| 1990 | 5.284 | 83 |
| 1991 | 5.367 | 83 |
| 1992 | 5.450 | 81 |
| 1993 | 5.531 | 80 |
| 1994 | 5.611 | 80 |
| 1995 | 5.691 | 78 |
| 1996 | 5.769 | 78 |
| 1997 | 5.847 | 78 |
| 1998 | 5.925 | 78 |
| 1999 | 6.003 | 78 |
| 2000 | 6.080 | 77 |
| 2001 (prel) | 6.157 | 77 |

*Total at mid-year.
Source: U.S. Bureau of the Census, *International Data Base*, electronic database, Suitland, MD, updated 10 May 2000.

*Ann Hwang*

Twenty years after it was recognized as a new disease, AIDS has claimed the lives of almost 25 million people—nearly equivalent to the population of Venezuela.[1] About 40 million more are living with HIV, the virus that causes AIDS. In 2001 alone, 5 million people became infected with the virus and 3 million died.[2] (See Figures 1 and 2.)

Sub-Saharan Africa remains the epidemic's epicenter: one tenth of the world lives there, but they account for nearly three quarters of the world's HIV infections.[3] AIDS is now that continent's leading cause of death.[4] Double-digit infection rates in many southern African countries have lowered life expectancy by 15 years, and in four countries—Botswana, Malawi, Mozambique, and Swaziland—people on average can now expect to die before they turn 40.[5] AIDS is claiming the lives of the continent's teachers, doctors, farmers, workers, and parents. As it does, it not only erases decades of social and economic progress but jeopardizes future growth. Some countries could lose more than 20 percent of their gross domestic product by 2020 due to the effect of AIDS on their work force and productivity.[6]

While infection rates elsewhere have not reached the catastrophic levels found in sub-Saharan Africa, the pace of the pandemic's spread is alarming. In Eastern Europe and Central Asia, the number of infections jumped 33 percent in 2001—from 750,000 to 1 million—fueled largely by the use of injection drugs.[7]

Asia—home to half the world—could become another disease epicenter. In a number of Indian states, more than 3 percent of the population is infected, a level that could spark an explosive disease spread.[8] Similar hot spots are found in China, where HIV is spreading through injection drug use, sexual contact, and, at least in the central provinces, unsanitary blood-selling practices. Some villages where blood-selling was common now have infection rates above 25 percent.[9]

In industrial and developing countries alike, discrimination compounds the suffering of people living with HIV/AIDS. Infected individuals have been fired from their jobs, disowned by their families, and even forcibly sterilized. A survey of 121 countries found that only 21 nations—representing 16 percent of the world's population—have specific laws to protect HIV-positive individuals from discrimination.[10]

In 1984, U.S. Health and Human Services Secretary Margaret Heckler predicted, "There will be a vaccine in a very few years and a cure for AIDS before 1990."[11] Though anti-retroviral therapy has prolonged the lives of many of those infected with HIV, there is still no cure. The therapies themselves have dangerous side effects, such as nerve damage and heart disease. And as HIV mutates, it can evade the drugs' effects and become resistant to treatment. Researchers from the Rand Corporation and the University of California at San Diego recently estimated that half of the HIV patients in the United States have a virus that is resistant to at least one anti-retroviral drug.[12]

In developing countries, where 95 percent of HIV-infected people live, anti-retroviral drugs are nearly impossible to obtain.[13] In sub-Saharan Africa, for example, only 30,000 people—one tenth of 1 percent of those infected—receive the triple anti-retroviral therapy recommended to combat HIV.[14] Despite opposition from pharmaceutical companies, some companies and countries are manufacturing generic versions of anti-retroviral drugs at a fraction of the price of the patented versions. South Africa's Treatment Action Campaign successfully sued the government to increase access to nevirapine, a drug that prevents the transmission of HIV from mother to child.[15]

But even deeply discounted drugs will likely be beyond the reach of most developing countries. And help from the industrial world may be slow in arriving. In April 2001, U.N. Secretary-General Kofi Annan announced the creation of a global fund to combat AIDS, tuberculosis, and malaria. The fund aimed to raise $7–10 billion, but by year's end had received only $2 billion in pledges.[16] And after September 11th, the U.S. Congress slashed its contribution to the new fund from nearly $1 billion to only $200 million—less than a dollar per American.[17]

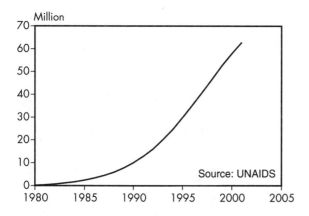

Figure 1: Estimates of Cumulative HIV Infections
Worldwide, 1980–2001

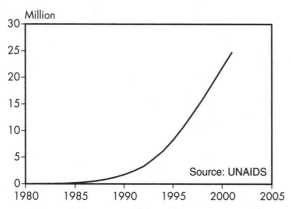

Figure 2: Estimates of Cumulative AIDS Deaths
Worldwide, 1980–2001

## Cumulative HIV Infections and AIDS Deaths Worldwide, 1980–2001

| Year | HIV Infections | AIDS Deaths |
|------|----------------|-------------|
| | (million) | |
| 1980 | 0.1 | 0.0 |
| 1981 | 0.3 | 0.0 |
| 1982 | 0.7 | 0.0 |
| 1983 | 1.2 | 0.0 |
| 1984 | 1.7 | 0.1 |
| 1985 | 2.4 | 0.2 |
| 1986 | 3.4 | 0.3 |
| 1987 | 4.5 | 0.5 |
| 1988 | 5.9 | 0.8 |
| 1989 | 7.8 | 1.2 |
| 1990 | 10.0 | 1.7 |
| 1991 | 12.8 | 2.4 |
| 1992 | 16.1 | 3.3 |
| 1993 | 20.1 | 4.7 |
| 1994 | 24.5 | 6.2 |
| 1995 | 29.8 | 8.2 |
| 1996 | 35.3 | 10.6 |
| 1997 | 40.9 | 13.2 |
| 1998 | 46.6 | 15.9 |
| 1999 | 52.6 | 18.8 |
| 2000 | 57.9 | 21.8 |
| 2001 (prel) | 62.9 | 24.8 |

Sources: UNAIDS, *AIDS Epidemic Update: December 2000* and *2001*; Neff Walker, UNAIDS, 20 March 2000.

# Military Trends

WWW.UNAUSA.ORG

Number of Violent Conflicts Declines

Peacekeeping Expenditures Rise Again

*Michael Renner*

The number of wars worldwide stood at 31 in 2001, down from 35 the previous year, according to AKUF, a conflict research group at the University of Hamburg.[1] (See Figure 1.) In addition, there were 15 "armed conflicts" active in 2001 that were not of sufficient severity to meet AKUF's criteria for war. Combining these two categories, the total number of violent clashes declined slightly—from 47 in 2000 to 46.[2]

The war between Ethiopia and Eritrea ended, and violence in Laos, Chiapas (Mexico), and Nigeria's oil-rich Niger delta subsided.[3] But three conflicts began during 2001: the war against the Taliban regime and the Al Qaeda network in Afghanistan, separatist violence by Albanians in Macedonia, and fighting between Christian and Muslim militias in Nigeria.[4]

Links: pp. 68, 96, 162

The significant decline in the number of conflicts during the 1990s is matched by a decline in the "magnitude" of violence. (The Center for International Development and Conflict Management (CIDCM) at the University of Maryland rates each conflict according to the number of deaths, dislocations, and physical damage wrought.)[5] Likewise, the proportion of countries involved in violent confrontations declined. In 1999, 18 percent of all states were at war, down from 33 percent in 1991.[6]

The September 11th terrorist attacks and the war in Afghanistan overshadowed virtually all other conflicts, and "anti-terrorism" strongly tinted the portrayal and public perception of a number of struggles, including the Israeli-Palestinian confrontation, Russia's fight against Chechen rebels, and the Indian-Pakistani standoff over Kashmir.

Most of the current conflicts are taking place in sub-Saharan Africa, the Middle East, and portions of Asia.[7] And CIDCM finds that countries in these regions "are at serious risk of armed conflict and political instability for the foreseeable future"—mostly because they lack stable and democratic institutions, suffer from a lack of resources, and have limited capacity to address ethnic and other disputes.[8]

It is becoming harder and harder to define and categorize violent conflicts, and not only because information about battles, tactics, motivations, and victims is spotty or unreliable. Armed forces are splintering in many countries even as private or semi-private security forces of various stripes multiply. And violent conflict is often not driven by ideology or the quest for government power but by the motivation to plunder lucrative resources such as diamonds, minerals, oil, and timber. Altogether, about a quarter of the armed conflicts waged during 2000 had a strong resource dimension.[9]

Different definitions and empirical methods among peace research groups lead to somewhat different results, although there is agreement on the broad, overall trends.[10] (See Figure 2.) Of 111 conflicts recorded by the researchers at the Uppsala Conflict Data Project during 1989–2000, 104 were internal (including 9 in which there was also foreign intervention).[11] Only 7 conflicts were interstate wars.[12]

Conflict researchers at the Heidelberg Institute for International Conflict Research in Germany (known as HIIK) cast a wider net than AKUF and the Uppsala group in their assessments of worldwide conflicts. HIIK reports that the number of political conflicts in the world has climbed fairly steadily from 108 in 1992 to 155 in 2001.[13] On the positive side, just 38 of the 155 conflicts were carried out by violent means.[14] (See Figure 3.) And HIIK finds that in more than one third of the conflicts active in 2001, negotiations and other means helped dampen the disputes.[15]

The overall conflict trends since 1990 are encouraging. But taken as a whole, the past century was extraordinarily violent. Milton Leitenberg of the University of Maryland estimates that from 1945 to 2000, some 50–51 million people were killed in wars and other violent conflicts.[16] For the entire twentieth century, he estimates 130–142 million war-related deaths, and a chilling 214–226 million if government killings in non-war situations are included.[17]

Figure 1: Wars and Armed Conflicts, 1950–2001

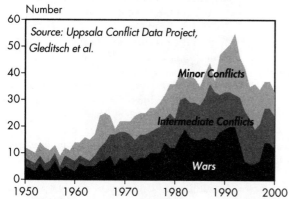

Figure 2: Wars and Intermediate and Minor Conflicts, 1950–2000

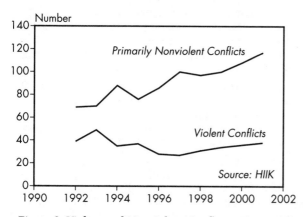

Figure 3: Violent and Nonviolent Conflicts, 1992–2001

## Wars and Armed Conflicts, 1950–2001

| Year | Wars | Wars and Armed Conflicts |
|------|------|--------------------------|
|      |      | (number) |
| 1950 | 12 | |
| 1955 | 14 | |
| 1960 | 10 | |
| 1965 | 27 | |
| 1970 | 30 | |
| 1971 | 30 | |
| 1972 | 29 | |
| 1973 | 29 | |
| 1974 | 29 | |
| 1975 | 34 | |
| 1976 | 33 | |
| 1977 | 35 | |
| 1978 | 36 | |
| 1979 | 37 | |
| 1980 | 36 | |
| 1981 | 37 | |
| 1982 | 39 | |
| 1983 | 39 | |
| 1984 | 40 | |
| 1985 | 40 | |
| 1986 | 42 | |
| 1987 | 43 | |
| 1988 | 44 | |
| 1989 | 42 | |
| 1990 | 48 | |
| 1991 | 50 | |
| 1992 | 51 | |
| 1993 | 45 | 62 |
| 1994 | 41 | 58 |
| 1995 | 36 | 51 |
| 1996 | 31 | 49 |
| 1997 | 29 | 47 |
| 1998 | 32 | 49 |
| 1999 | 34 | 48 |
| 2000 | 35 | 47 |
| 2001(prel) | 31 | 46 |

Source: Arbeitsgemeinschaft Kriegsursachenforschung, Institute for Political Science, University of Hamburg.

*Michael Renner*

Expenditures for United Nations peacekeeping operations are expected to continue their rapid upswing, growing from $2.6 billion for the July 2000–June 2001 period to an estimated $2.7–3 billion for July 2001 to June 2002.[1] (See Figure 1.) This means that peacekeeping spending is now edging toward the peak budgets of the mid-1990s.

More than 47,000 soldiers, military observers, and civilian police served in 15 peacekeeping missions active at the end of 2001, up 24 percent from about 38,000 a year earlier.[2] (See Figure 2.) The missions were supported by 12,126 local and international civilian personnel.[3] (In addition to peacekeeping and observer operations, the United Nations also maintained 13 small political and peace-building missions involving about 600 mostly civilian staff; one of these has been working in Afghanistan since 1993.)[4] Since the inception of peacekeeping operations in 1948, a total of 1,706 peacekeepers have died in the line of duty.[5]

Link: p. 94

Ninety countries contributed personnel to the U.N. missions during 2001.[6] Bangladesh and Pakistan scaled up their involvement dramatically; these two countries together currently account for about one fifth of all deployed peacekeepers.[7] Nigeria, India, Jordan, Ghana, Kenya, and Australia are also major contributors. Rounded out by Ukraine and Portugal, the leading 10 sources of personnel provided 58 percent of the total.[8] The five permanent members of the Security Council, by comparison, kept their involvement limited to about 6 percent.[9]

No new missions were initiated or authorized during 2001. On 27 March and 15 December 2001, the United States vetoed resolutions before the U.N. Security Council to establish a U.N. observer force to protect Palestinian civilians in the West Bank and Gaza Strip and to send monitors to help prevent further Israeli-Palestinian violence.[10] The vetoes followed similar votes in December 2000.[11]

U.N. peacekeeping activities and expenditures continued to be dominated by just three operations.[12] About 17,000 peacekeepers—more than a third of the total—are stationed in Sierra Leone alone, where the United Nations is trying to end a decade-long conflict revolving around lucrative diamond resources.[13] Some 8,500 peacekeepers are in East Timor, and about 4,500 in Kosovo.[14] But sizable deployments are also found in southern Lebanon, at the border separating Ethiopia and Eritrea, and in the Democratic Republic of Congo.[15]

Other missions continue at the India-Pakistan border (since 1949), in Cyprus (1964), on the Golan Heights separating Israel and Syria (1974), at the Iraq-Kuwait border (1991), in Western Sahara (1991), in Georgia (1993), in Bosnia (1995), and on the Prevlaka peninsula between Croatia and Serbia (1996).[16]

As of the end of October 2001, U.N. members owed the organization $1.9 billion for peacekeeping operations.[17] (See Figure 3.) The United States accounts for 41 percent of the total unpaid dues, or $787 million.[18] Following payment of some long-standing arrears, this is a significantly lower share than in recent years.[19] With these payments, the United Nations hopes that "for the first time in many years [it] might have a secure basis with which to do business."[20]

In addition to U.N. peacekeeping operations, some three dozen additional missions are being carried out by regional or military organizations, such as NATO, the Organization for Security and Co-operation in Europe, and the Economic Community of West African States, or by ad hoc coalitions of states. Many of them are very small. By far the largest are NATO-led operations in Bosnia, Kosovo, and Macedonia. Together, these Balkan missions deploy about 60,000 soldiers and cost an estimated $8–9 billion annually.[21]

In December 2001, the U.N. Security Council endorsed creation of a British-led International Security Assistance Force to ensure security in Kabul, Afghanistan's capital, following ouster of the Taliban.[22] A force of up to 5,000 soldiers was authorized for a six-month period.

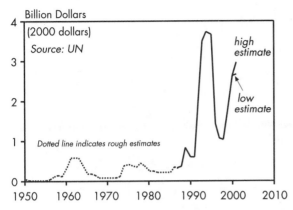

Figure 1: U.N. Peacekeeping Expenditures, 1950–2001

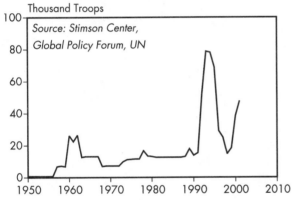

Figure 2: U.N. Peacekeeping Personnel, 1950–2001

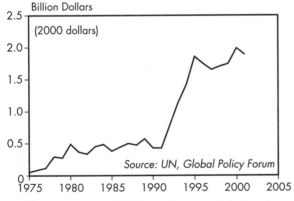

Figure 3: Arrears of U.N. Members for Peacekeeping
Expenses, 1975–2001

## U.N. Peacekeeping Expenditures, 1986–2001

| Year | Expenditure |
|------|-------------|
|  | (bill. 2000 dollars) |
| 1986 | 0.344 |
| 1987 | 0.331 |
| 1988 | 0.355 |
| 1989 | 0.815 |
| 1990 | 0.573 |
| 1991 | 0.585 |
| 1992 | 2.058 |
| 1993 | 3.480 |
| 1994 | 3.724 |
| 1995 | 3.668 |
| 1996* | 1.423 |
| 1997* | 1.039 |
| 1998* | 1.037 |
| 1999* | 1.683 |
| 2000* | 2.630 |
| 2001* (low) | 2.650 |
| (high) | 2.950 |

* July to June of following year.
Sources: U.N. Department of Peacekeeping
Operations; U.N. Department of Public
Information.

# PART TWO
## Special Features

# Environment Features

© DIGITAL VISION

# Farmland Quality Deteriorating

*Brian Halweil*

A substantial area of the world's farmland is degraded and getting more so, particularly in the developing world.[1] A recent analysis found that 10–20 percent of the world's 1.5 billion hectares of cropland—150–300 million hectares—suffers from some level of degradation.[2] Moderate, severe, or extreme degradation affects 7–14 percent, or 105–210 million hectares.[3] These estimates come from a reanalysis of data collected for the 1991 *Global Assessment of the Status of Human-Induced Soil Degradation* (GLASOD).[4]

Another recent survey of land degradation studies found that roughly one quarter of the farmland in the developing world suffers from degradation, and the pace of decline has accelerated in the past 50 years.[5] Compared with the industrial world's soils, the tropical soils of the developing world are older (they were not rejuvenated by the last glaciation), exposed to more severe weather, more often in hilly or mountainous areas, and require more careful management to avoid degradation.[6]

Links: pp. 26, 30, 34, 152

Farmland in arid areas—both rangeland and cropland—is particularly susceptible to degradation, because the low rainfall and sparse vegetation mean that soils and plants recover more slowly.[7] Over 70 percent of the world's rangelands—which cover 3.4 billion hectares worldwide and are found mostly in arid areas—suffers from moderate to very severe degradation as a result of overgrazing, changes in rainfall, and deforestation.[8]

Among the most common causes of farmland degradation are excessive tillage and removal of vegetation (including crops and forests), which leaves the soil exposed to rain and wind. Too many animals feeding on an area of land can also strip it of vegetation and expose it to erosion and other degradation. GLASOD attributes about 35 percent of human-induced degradation around the world to overgrazing and about 28 percent to other forms of agricultural mismanagement.[9] Inappropriate use of land not suited to agriculture, because it is too dry or steeply sloping, can also lead to degradation. A survey of Central American cropland found that nearly half is used inappropriately—more than 30 percent of the region's land is used for grazing, while only 15 percent is actually suited for pasture.[10]

While excessive use of fertilizers causes widespread damage to soils and waterways in wealthy nations, in the developing world farmland generally suffers from the depletion of nutrients as farmers continuously harvest crops without fertilizing or fallowing the land. Farmers in Central Africa lose 30–60 kilograms of nutrients (primarily nitrogen, phosphorus, and potassium) per hectare each year, a figure that climbs to above 60 kilograms in East Africa.[11] In Latin America and the Caribbean, the region's soils lose around 54 kilograms per hectare each year, with losses concentrated in Argentina and Brazil.[12]

Degradation undercuts food production and farm income, as the land supports smaller harvests and costs more to maintain.[13] (See Table 1.) Each year, some 5–8 million hectares of farmland go out of production as a result of degradation.[14] Worldwide, land degradation has reduced cumulative food production by an estimated 13 percent on cropland and 4 percent for pasture over the last half-century.[15] A study of West Africa found that child mortality was highest in areas with the highest soil degradation.[16]

Soil erosion is perhaps the most damaging form of farmland degradation, because it removes the foundation on which crops, wild plants, and other life subsist and because it takes hundreds of years for soils to rebuild. GLASOD suggested that erosion by water (when rain removes soil from fields) is the dominant form of degradation on all continents, present on half the world's degraded lands.[17] Wind erosion accounts for another 30 percent.[18]

In the United States, one of the few nations where erosion rates have been tracked for several decades, the rate of erosion has declined substantially since 1982, from 2.65 tons per hectare in 1982 to 1.8 tons in 1997.[19] Despite these improvements—largely attributed to greater adoption of reduced tillage practices and efforts to set aside highly erodible cropland—the nation still loses nearly 6 tons of soil for each ton of grain harvested.[20]

Salinization is the most common form of degradation on irrigated cropland. This buildup of salts, as excess irrigation water evaporates and concentrates toxic salts near the soil surface, can devastate yields, and often force the abandonment of irrigated land altogether. An estimated 47.7 million hectares of land worldwide—some 20 percent of the world's total irrigated land—are damaged by salinization, costing farmers roughly $11 billion each year in reduced harvests.[21]

Beyond the farm, degradation can damage water supplies, roads, and other infrastructure through soil erosion, runoff, flooding, and dam sedimentation.[22] At the global level, farmland degradation releases carbon dioxide from soils into the atmosphere and can fuel dust clouds and sandstorms that blow across continents and even oceans.[23] In extreme cases, soil degradation can prompt massive human movements; worldwide, desertification (land degradation in arid areas) could displace more than 135 million, and threatens the livelihoods of more than 1 billion people.[24]

Farmers can help reverse land degradation by improving fertilization practices, planting tree crops, and using cover crops (crops added to the rotation to protect the soil), green manures (crops that protect the soil and add nutrients), and other techniques that help protect and build soil.[25] Among the more promising trends is the rapid shift by some farmers to "no-till" practices, which involve planting seeds in the stubble of the previous crop rather than plowing each season, which can accelerate erosion.[26] Farmers are using no-till on 11 million hectares in Brazil, up from 1 million in 1991, and on 9.2 million hectares in Argentina, up from 100,000 hectares in 1990.[27] In Latin America, the technique has cut soil erosion by as much as 90 percent.[28]

## Table 1: Selected Examples of the Consequence of Farmland Degradation

| | |
|---|---|
| Reduced agricultural productivity | Degradation cut productivity by one third on half of India's soils. In wheat-rice cropping systems of the Pakistani and Indian Punjab, degradation more than cancelled yield-enhancing effects of 40 years of technological change. Yield reductions of 25–50 percent predicted in Argentina, Kenya, and Uruguay over next 20 years. |
| On-farm expenses | Nutrient depletion costs sub-Saharan Africa about 7 percent of agricultural production a year in terms of equivalent amounts of purchased fertilizer. Depletion amounts to $4 billion per year, much more than development assistance to African agriculture. In the early 1990s, on-site costs of soil degradation cost South Asia $9.8–11 billion each year—7 percent of agricultural GDP. |
| Salinization | Agricultural production threatened in virtually all the world's irrigated regions, particularly South and Southeast Asia. Share of land in Bangladesh affected by salinization nearly quadrupled since 1990—from 9 to 34 percent. In four villages in Uttar Pradesh, India, salinization and waterlogging reduced rice yield by 61 percent and wheat yield by 68 percent over 10 years. |
| Off-farm expenses (air pollution, road damage, water pollution, desertification) | For 200 major dams worldwide, buildup of soil—sedimentation—costs $4 billion a year in reduced irrigation and hydropower and in additional maintenance. Sandstorms from Inner Mongolia darken the air in Beijing and 20 other major cities in northern China, while dust storms from Africa blamed for spreading a soil-borne fungus to Caribbean coral reefs. U.S. public benefits from erosion reduction, including higher farm productivity, reduced cleanup costs, and higher quality of water bodies, conservatively estimated at $1.4 billion a year. |

*Source:* See endnote 13.

# Forest Loss Unchecked

*Janet N. Abramovitz*

In 2001, the U.N. Food and Agriculture Organization (FAO) completed its latest Global Forest Resources Assessment and reported that during the 1990s "the world's natural forests continued to be converted to other land uses at a very high rate."[1] FAO estimated that at least 4.2 percent of the forest cover that stood in 1990 was gone by the end of the decade.[2]

FAO found that 161 million hectares of natural forest were lost during the decade, and 152 million hectares of the loss occurred in the tropics.[3] Only a small amont of natural forest loss was offset by regrowth—just 36 million hectares during the decade.[4] Only 10 million hectares of that growth was in the tropics.[5]

About half the Earth's original forest cover is gone, and another 30 percent has been degraded or fragmented, according to reliable estimates by the World Resources Institute.[6]

During the last decade, the vast majority of the total forest cover loss—that is, loss of natural forests offset by regrowth or expanded plantation cover—reported by FAO occurred in just eight countries: Brazil, China, Indonesia, Sudan, Zambia, Mexico, the Democratic Republic of Congo, and Myanmar (formerly Burma).[7] The forest cover of these eight together declined by 89.2 million hectares.[8] Brazil alone lost 23.1 million hectares, China 18.1 million hectares, and Indonesia 13.1 million hectares.[9]

*Links:* pp. 70, 108

In 2000, the world's forest cover stood at 3,869 million hectares, about 95 percent of which was natural forest and the rest plantation forest.[10] Ten countries contain two thirds of the world's total forest cover: the Russian Federation, Brazil, Canada, the United States, China, Australia, the Democratic Republic of Congo, Indonesia, Angola, and Peru.[11] In terms of just natural forest, South America holds 24 percent of the total, the Russian Federation has 23 percent, and Africa, 17 percent.[12] (See Table 1.)

Today, 57 percent of the world's forests are tropical, 33 percent boreal, 11 percent temperate, and 9 percent subtropical.[13] Most tropical moist forests are in South America (58 percent), while 24 percent are in Africa, and 17 percent in Asia.[14] Africa holds the largest share of tropical and subtropical dry forest (36 percent), while South America holds 30 percent of this forest type, and Asia has 21 percent.[15]

Tree plantations expanded by 31 million hectares during the decade—and half of that came at the expense of natural forests that were removed to make way for the plantations.[16] As of 2000, there were 187 million hectares of tree plantations.[17] The lion's share—some 62 percent—is found in Asia, with China and India in the lead.[18] Plantations now account for 21 percent of Asia's forest cover.[19] At least half (48 percent) of the world's plantations are for industrial uses like lumber and paper.[20] Over a quarter (26 percent) are for fuel or to protect soil and water.[21] (The purpose of the remainder was not recorded.)[22]

Many nations lost a high portion of their forests during the last decade. Eighteen nations lost 20 percent or more of their forest cover, while another 16 lost 10–19 percent.[23] Most of the highest losses were recorded in Africa: Rwanda and Burundi each lost 39 percent, and Côte d'Ivoire, Sierra Leone, and Niger each lost about a third.[24] Another troubled African nation, Liberia, recorded a 20-percent loss, although recent reports exposing widespread illegal logging may mean that this figure is low.[25] El Salvador, Nicaragua, and Belize each lost between a quarter and a third of their forest cover during the 1990s, while Guatemala lost about 17 percent.[26] (Some of the consequences of high deforestation were seen when Hurricane Mitch devastated Central America in 1998.)[27] High-loss nations in Asia include Nepal, Sri Lanka, Pakistan, Myanmar, the Philippines, Indonesia, and Malaysia.[28]

In order to manage forests more sustainably, forest monitoring must be improved, along with the way that official forest data are reported. The FAO defines "deforestation" as a permanent conversion of forest to other uses (such as agriculture) or a long-term (10 or more years) reduction of canopy cover to less than 10 percent.[29] Thus, a forest can be denuded or highly fragmented for nine years and still be counted as forest. This highlights the difference between the official definition of defor-

## Table 1: Natural and Plantation Forest Area, by Region, 2000

| Region | Total Land Area | Natural Forest Area | Share of World's Natural Forest Area | Plantation Area | Share of World's Plantation Area |
|---|---|---|---|---|---|
| | (million hectares) | | (percent) | (million hectares) | (percent) |
| Africa | 2,978 | 642 | 17 | 8 | 4 |
| Asia | 3,085 | 432 | 12 | 116 | 62 |
| Oceania | 849 | 194 | 5 | 3 | 2 |
| Europe | 571 | 173 | 5 | 14 | 7 |
| Russian Federation[1] | 1,689 | 834 | 23 | 17 | 9 |
| North and Central America | 2,137 | 532 | 14 | 18 | 10 |
| South America | 1,755 | 875 | 24 | 10 | 6 |
| World | 13,064 | 3,682 | 100 | 187 | 100 |

[1]Included within Europe in original FAO data.

Source: U.N Food and Agriculture Organization, *State of the World's Forests 2001* (Rome: 2001), pp. 37, 41, 152.

estation and a more commonly understood use of the term. The inclusion of plantations in estimates of global forest cover (even when those plantations replaced natural forest) can also lead to a distorted understanding of forest trends. For its latest forest resources assessment, FAO revised many of the methodologies and definitions used, and thus cautions that the latest numbers cannot be compared with those from earlier assessments.[30]

Better monitoring of the forests through use of satellite data and on-the-ground monitoring is also needed. As FAO itself reveals, there are significant problems in the quality and comparability of the data it collects from individual countries. The lack of on-the-ground forest inventories and scanty satellite monitoring are major barriers. Three quarters of developing countries have either never carried out a forest inventory or have done only one, making accurate assessments of changes over time nearly impossible.[31] Forest management and monitoring are chronically understaffed and underfunded in many nations.

Independent monitoring groups play an important role in identifying forest conditions and assessing the veracity of official data. For example, Global Forest Watch (GFW) and its network of in-country partner organizations have undertaken in-depth studies of several countries, including Canada, Cameroon, Gabon, Indonesia, and Russia.[32] Forest Watch Indonesia and GFW reported in 2002 that Indonesia lost 40 percent of its forests since 1950, and in the last 20 years the rate of loss has doubled to about 2 million hectares per year.[33] Other groups, including Global Witness, Greenpeace, Telepak, and the Environmental Investigation Agency, are also tracking illegal forest destruction.[34]

In 2001 the U.N. Environment Programme (UNEP), in collaboration with NASA and the U.S. Geological Survey, produced an assessment of the world's remaining closed forests, which it defined as virgin, old growth, or naturally regenerated forests with a canopy density of greater than 40 percent. It reported that in 1995 this category covered about 2.87 billion hectares.[35] Together, Russia, Canada, and Brazil had 49 percent of this total.[36] UNEP noted that about half of the remaining closed forests are "more or less intact," but echoed the assessment of many that "the remaining forests [are] very fragmented and under high pressure."[37]

# Freshwater Species at Increasing Risk     *Sandra Postel*

Species that depend on rivers, lakes, wetlands, and other freshwater environments for a major portion of their lifecycle are being imperiled and extinguished at an alarming pace. The principal culprit is the destruction of freshwater habitats by dams, river diversions, and pollution, along with the introduction of non-native species. Because communities of freshwater species perform valuable ecological services—filtering and cleansing water supplies, mitigating floods and droughts, and delivering nutrients to the sea, for example—stepped-up efforts to stem the tide of biological decline are needed urgently.

A comprehensive global assessment of freshwater biodiversity is not possible because of the lack of data for most countries. But researchers estimate that at least 20 percent of the world's 10,000 freshwater fish species are now endangered, are threatened with extinction, or have already gone extinct.[1] A significant but unknown share of mussels, amphibians, aquatic insects, and other species that depend on fresh water are also at risk. Many species may be lost even before they are found or named: indeed, scientists have been describing about 300 new freshwater species each year.[2]

Link: p. 24

In North America, at least 123 species of freshwater fish, mollusks, crayfish, and amphibians have become extinct since 1900.[3] Biologists Anthony Ricciardi and Joseph Rasmussen estimate that in recent decades North American freshwater animal species have been extinguished at an average rate of half a percent per decade.[4] They project, moreover, that this will increase in the near future to 3.7 percent a decade—about five times greater than the projected extinction rate for North American terrestrial animal species.[5] In fact, the relative rate of loss of North American freshwater species is comparable to that of species in tropical rainforests.[6]

The United States stands out as a global center of freshwater biodiversity. The nation ranks first in the world in the number of known species of freshwater mussels, snails, and salamanders, as well as three important insect groups—caddisflies, mayflies, and stoneflies.[7] U.S. waters are home to 300 species of freshwater mussels—29 percent of those known worldwide—and nearly twice as many as live in Europe, Africa, India, and China combined.[8] With approximately 800 species of freshwater fish, the United States ranks seventh in freshwater fish diversity globally but has by far the most diverse assemblage of fishes of any temperate country.[9]

In the most comprehensive survey to date of the conservation status of U.S. plant and animal species, researchers with The Nature Conservancy and the Association for Biodiversity Information found that of 14 major groups of organisms, the 5 with the greatest share of species at risk were all animals that depend on freshwater systems for all or part of their lifecycle.[10] (See Table 1.) An astonishing 69 percent of U.S. freshwater mussels are to some degree at risk of extinction or are already extinct —compared with 33 percent of flowering plants, 16 percent of mammals, and 14 percent of birds.[11]

Although no comparable surveys exist for most of the rest of the world, the prognosis for freshwater life is not good. Swedish scientists Mats Dynesius and Christer Nilsson have found that 77 percent of the 139 largest river systems in the United States, Canada, Europe, and the former Soviet Union—essentially the northern third of the world—are moderately to strongly altered by dams, reservoirs, diversions, and irrigation projects.[12] Worldwide, the number of large dams (those at least 15 meters high) stood at 5,000 in 1950, and three quarters of these were in North America, Europe, and other industrial regions.[13] By 2000, there were more than 45,000 large dams and they were spread among more than 140 countries.[14]

Most new dam construction and major river diversions are occurring in developing countries as they strive to increase irrigation, water supplies, and hydroelectric power, much as industrial countries did before them. Consequently, the rich diversity of freshwater life in tropical Asia, Africa, and Latin America will come under increasing pressure. The

## Table 1: Risk Status of U.S. Animal Species Dependent on Freshwater Ecosystems

| Animal Group | Total Number of Species | Share that is Extinct, Critically Imperiled, Imperiled, or Vulnerable |
|---|---|---|
| | | (percent) |
| Freshwater Mussels | 292 | 69 |
| Crayfishes | 322 | 51 |
| Stoneflies | 606 | 43 |
| Freshwater Fishes | 799 | 37 |
| Amphibians | 231 | 36 |

Source: See endnote 10.

Amazon basin alone harbors more than 2,000 species of freshwater fish—about one in five of those known worldwide—and scientists estimate that 90 percent are found nowhere else.[15] With more than 70 dams planned for Brazil's Amazonian region alone, a good portion of these species are likely to be threatened.[16]

Asia also has a diverse array of freshwater species coming under increasing threat from habitat destruction. Indonesia has at least 1,200 freshwater fish species, China more than 700, and Thailand more than 500.[17] Asian rivers are home to three of the world's five species of true river dolphins—those that never enter the sea—and all three are endangered.[18] Tropical Asia also harbors the world's richest assemblage of freshwater turtles, as well as 8 of the world's 23 crocodilian species.[19] All 8 are now endangered.[20]

The ecology of Asian rivers is driven largely by the monsoons, which create high and low river flows at fairly predictable times of the year. The organisms that inhabit these rivers have adapted and keyed their lifecycles to this flow pattern over time. Dams not only block many of them from migrating up or down river, they smooth out the flow of rivers, thereby eliminating habitats and environmental cues that various species need to complete their lifecycles. They also disconnect rivers from their floodplains, which many species rely on for breeding and feeding.

Combined with pollution, watershed degradation, and the introduction of non-native species, additional dam construction will place a greater proportion of Asian freshwater species at risk. In Southeast Asia, the Mekong Commission has identified a dozen sites for dams on the Mekong River in Laos, Thailand, and Cambodia.[21] Dam construction continues in China, which already has nearly half of the world's large dams.[22]

Finally, the algae, fungi, worms, and other species that live in freshwater environments are also at risk from the alteration of aquatic habitats. Globally, more than 100,000 species of invertebrates are estimated to live in freshwater sediments, along with 10,000 species of algae and more than 20,000 species of protozoa and bacteria.[23] These tiny sediment-dwellers help maintain water quality, decompose organic matter, produce food for animals higher in the food chain, and perform other critical functions. Scientists have found them to be very sensitive to changes in water levels, flow magnitudes, and other hydrologic alterations.[24]

Protecting the valuable ecosystem services upon which society depends requires conserving the unique assemblages of species that perform this work. This, in turn, requires building habitat protection into the management and use of rivers. A guiding principle now gaining ground is that of a freshwater "reserve"—the notion that ecosystems should be allocated the quantity, quality, and timing of freshwater flows needed to maintain their health and functioning.[25] South Africa is pioneering the implementation of this principle following passage in 1998 of a new water act that calls for the establishment of ecological reserves for its rivers.[26]

# Transboundary Parks Become Popular     *Lisa Mastny*

In recent years, "transboundary parks" have become an important tool for conserving the planet's biodiversity and promoting regional stability. These parks are formed when neighboring countries agree to link and jointly manage national parks, wildlife reserves, or other protected areas that are adjacent but lie on opposite sides of a shared border.[1]

The earliest effort to unify two adjoining parks dates to after World War I, when the 1925 Cracow Protocol called for the creation of twin national parks along the then-disputed Czech-Polish border.[2] Today, transboundary parks—also known as peace parks—are found on six continents, from South America to Asia.

*Links*: pp. 94, 104

In some cases, the level of cooperation between neighbors is highly formal: in 1932, when the United States and Canada created North America's first transboundary park, the Waterton-Glacier International Peace Park, they signed an international treaty.[3] Poland and Belarus, in contrast, have yet to forge diplomatic ties between their neighboring parks—Bialowieza and Belovezhskaya Pushcha—although they cooperate scientifically by exchanging plants and wildlife.[4] And India and Bhutan coordinate only anti-poaching efforts in their adjacent Manas parks.[5]

Opportunities for cross-border conservation are growing as countries designate new protected areas along their boundaries.[6] Researcher Dorothy Zbicz estimates that in 1988, in only 59 sites worldwide did adjoining protected areas lay on opposite sides of a national border.[7] By 2001, the figure had nearly tripled, to 169 sites.[8] Some degree of transboundary cooperation already occurs at many of these locations, though typically at the lowest levels.[9]

The sites straddle about a third of the world's more than 300 international boundaries and are distributed among 113 different countries.[10] The majority of the sites span just two countries, but as many as 31 cover three nations.[11] Most are located in Europe.[12] (See Table 1.) Altogether, these transboundary areas account for more than 10 percent of the currently protected land area worldwide.[13]

In addition to the 169 sites, there are at least as many border locations where adjoining protected areas do not yet exist but could be established—creating hundreds of opportunities for future cross-border conservation.[14] These include places where a park or reserve is found on only one side of the border, or on neither side, but where protection is still viable.[15]

By establishing transboundary parks, conservationists hope to reconnect single ecosystems that have been artificially severed by political boundaries. By one estimate, more than half of all international borders were drawn up arbitrarily by just six colonial powers, typically as an outcome of war or political compromise.[16] Many of these borders bisect continuous deserts, forests, and watersheds, greatly increasing the political challenge of managing these areas.[17] (The habitat of Africa's endangered mountain gorilla, for instance, is in a war-torn region shared by Rwanda, Uganda, and the Democratic Republic of Congo.)[18]

Because of their large size, transboundary parks may be more effective than national parks at stemming species extinctions and protecting valuable ecological processes.[19] For instance, they may be better able to support a more diverse gene pool for an animal or plant population, or to encompass the range required for large mammals like elephant or buffalo.[20] Transboundary parks can also serve as important wildlife corridors, recreating ancient migration paths on land or water.[21]

There are administrative benefits as well. Often, park officials do not communicate or coordinate activities with their cross-border counterparts, though they may face similar challenges.[22] By collaborating, parks can maximize efficiencies of scale and avoid duplication—sharing the costs for research, education, training, or equipment, for instance, or jointly combating illegal logging or wildfires.[23]

The very process of linking protected areas can foster dialogue among long-distrustful neighbors.[24] By one estimate, more than half of all countries share borders that are ill defined and contested.[25] By collaborating through "peace parks," governments can boost regional security and build understanding and reconciliation among communities and institutions.[26]

Arthur H. Westing, an expert on transboundary conservation, argues that peace parks can boost political security in three general ways: by reinforcing relations among friendly neighbors, by easing tensions among sparring neighbors, or by facilitating reunification of divided countries, such as the two Koreas.[27] Already, provisions for peace parks have been incorporated into the treaty resolving the 1998 territorial dispute between Peru and Ecuador, and are also being used in negotiations between Israel and its neighbors.[28]

The creation of transboundary parks can also boost the welfare of local people living in border areas, provided they are active participants in any revenue-generating activities.[29] Communities living in and around southern

Africa's newly created peace parks, for instance, hope to capitalize on joint tourism activities.[30] Participating regions can also benefit from the cooperative management of shared resources, such as watersheds or fisheries.[31]

But transboundary conservation still faces many obstacles. Neighboring countries may share similar ecosystems, yet they often have quite different cultural and political values, forms of governance, and levels of stability.[32] Their adjacent parks may vary in infrastructure and in some more localized problems.[33] And the cost of unifying parks can be high: funds may be needed for land purchases or leases, removal of fencing, staff, counter poaching, wildlife reintroductions, or community development projects.[34] In most cases, however, the benefits of transboundary parks to nature and society will outweigh these costs.

## Table 1: Selected Opportunities for Transboundary Conservation, by Region

Europe (64 sites with adjoining protected areas)
At least 50 formal transboundary parks exist, many of which straddle the former Iron Curtain. In February 2000, Albania, Greece, and Macedonia created southeastern Europe's first transboundary park, the shared Prespa Park wetland area.

Africa (36 sites)
The continent's first peace park, the Kgalagadi Transfrontier Park shared by South Africa and Botswana, opened in May 2000. Four subsequent parks also span South Africa and its neighbors. Efforts to link mountain gorilla reserves in Uganda, Rwanda, and Democratic Republic of Congo remain impeded by ongoing conflict.

Asia (30 sites)
In September 1998, Russia, China, Mongolia, and Kazakhstan announced cooperation in conserving the shared Altai Mountains area. In May 1999, Nepal and India agreed to join several parks to create a single wildlife corridor. South Korea supports formally protecting parts of Korea's largely pristine demilitarized zone, though North Korea does not.

Central and South America (29 sites)
The region's first transboundary park, La Amistad, was created in 1982 to promote peace between Costa Rica and Panama. In 1988, Costa Rica and Nicaragua linked 51 different protected areas through their Si-A-Paz project. A proposed Meso-American Biological Corridor could link existing protected areas in eight countries.

North America (10 sites)
The region's first peace park, Waterton-Glacier, was established on the U.S.–Canadian border in 1932. In 1997, the United States and Mexico agreed to link adjoining parks in the Rio Grande valley. Since 1990, Russia and the United States have considered creating a shared park bridging the Bering Strait, although the idea has faced political opposition.

Source: See endnote 13.

# Semiconductors Have Hidden Costs    *Ann Hwang*

In 2001, there were 60 million transistors produced for every man, woman, and child on Earth.[1] These tiny components are used to build semiconductor chips, the brains behind many electronic devices: computers, of course, but also cars, microwaves, cellular phones, vending machines, and even musical greeting cards. By 2010, transistors will become even more pervasive, with 1 billion expected to be produced per person.[2]

The semiconductor industry has grown explosively in the past two decades. In 1982, annual sales of semiconductor chips totaled $14 billion.[3] By 2000, sales exceeded $200 billion.[4]

*Links*: pp. 82, 84, 112

Although the market contracted to an estimated $140 billion in 2001, the industry was showing signs of recovery in early 2002.[5]

As semiconductor technology has advanced, chips have become smaller, cheaper, and more numerous. In 1972, one megabyte of semiconductor memory cost $550,000; today, it costs only a few dollars.[6] (Most handheld personal organizers now contain eight megabytes of memory.) The semiconductor industry has the capacity to produce 69 million wafers a year—each wafer holds anywhere from a handful to thousands of chips—with plants operating at 64–80 percent of capacity in 2001.[7]

The production of these high-tech marvels requires the relatively low-tech ingredients of human labor and chemicals—lots of them. Computer chips are created from silicon that has been refined, molded, cut, and polished (often with strong acids) into a thin wafer. The electronic circuits that carry out the chip's functions are etched into the wafer's surface in a process akin to stenciling: one set of chemicals is applied to mask parts of the surface, another to etch the exposed surface, and a third to remove the first set.

The semiconductor industry is one of the most chemically intensive ever known.[8] A single plant may use 500–1,000 chemicals.[9] Manufacturing each silicon wafer not only requires tremendous amounts of chemical ingredients, it generates huge volumes of chemical waste. (See Table 1.) Santa Clara County, the birthplace of the semiconductor industry, now contains more U.S. Environmental Protection Agency (EPA) Superfund (toxic waste) sites than any other county in the nation.[10]

Workers in the "clean rooms" where chips are made handle these toxic chemicals every day. Clean rooms keep dust and other particles from spoiling the delicate silicon wafers, but are not necessarily clean for workers. (In the United States, the semiconductor industry employed 284,000 people in 1999; around the world, the work force may exceed 1 million.)[11] Women working in these rooms who handled reagents containing glycol ethers were found to have a 40-percent increase in their miscarriage rate compared with women without clean room exposure.[12] Although semiconductor manufacturers have since phased out glycol ethers, little research has been done on the other chemicals that clean room workers are exposed to.

To keep up in this fast-paced industry, companies may alter their manufacturing process without studying the long-term health and environmental effects of the new chemicals or processes. Another challenge for occupational health researchers and providers is that workers are exposed to mixtures of chemicals, and relatively little is known about whether exposures to mixtures rather than single chemicals can have unexpected health effects.

Efforts to fill some of these data gaps have at times met with reluctance, if not outright resistance, from the semiconductor industry. In 1998, the EPA funded and the California Department of Health Services agreed to conduct a study of cancer and birth defect rates among the state's semiconductor workers.[13] Despite the state's promise of confidentiality for workers and companies, the industry withdrew from the project at the last minute. Intel spokesman Tim Mohin famously declared, "To participate in a project like this would be like giving discovery to plaintiffs. I might as well take a gun and shoot myself."[14]

The threat of litigation is real. IBM and National Semiconductor are facing lawsuits in California, New York, and Scotland.[15] The

## Table 1: Resources Required and Waste Produced Per Six-Inch Semiconductor Wafer

| Resources Required | Waste Produced |
| --- | --- |
| 90 cubic meters of bulk gases | 11 kilograms of sodium hydroxide |
| 0.6 cubic meters of hazardous gases | 11,000 liters of waste water |
| 8,600 liters of water | 3 kilograms of hazardous waste |
| 9 kilograms of chemicals | |
| 285 kilowatt-hours of electricity | |

Note: Updated estimates unavailable. Plant and company data suggest improved recycling and decreased releases, but industry-wide efficiency could not be ascertained.
Source: Gordon Larabee, Texas Instruments, 1993, at <www.svtc.org/hightech_prod/larachart.htm>, viewed 30 September 2001.

plaintiffs allege that years of exposure to toxic chemicals caused cancer and birth defects. In January 2001, IBM settled out of court for an undisclosed amount with 15-year-old Zachary Ruffing, who was born with multiple birth defects.[16] Both of his parents had worked at an IBM plant in New York in the 1980s.

Even at the end of their life cycle, semiconductors continue to pose environmental challenges. Businesses and consumers now generate an almost continual turnover of electronic products. In some American businesses, the rule of thumb has been "one computer per user per year." The short life span and increased number of these products are fueling a growing waste crisis. Approximately 6 million tons of electronic waste were produced in the European Union in 1998.[17] This volume is expected to increase by at least 3–5 percent a year, or three times faster than the waste stream as a whole.[18] In the United States, more than 2.9 million tons of e-waste ended up in landfills in 1997, with the amount predicted to increase fourfold in the next few years.[19] But worse is yet to come: at least 315 million computers in the United States are predicted to become obsolete by 2004.[20]

Increased recycling is part of the solution. The National Safety Council estimates that in the United States in 1999, only 11 percent of discarded computers were recycled.[21] In 2001, both IBM and Hewlett Packard announced U.S.

recycling programs, which charge the consumer approximately $30 per computer.[22] In April, faced with landfills rapidly reaching capacity, the Japanese government enacted an Appliance Recycling Law requiring consumers to pay manufacturers a fee to recycle discarded appliances.[23] The law covers televisions, air conditioners, washing machines, and refrigerators, with computers to be added in the future. And in June, the European Council approved a directive requiring manufacturers of electronic equipment to pay for the recycling of their products.[24] (The European Parliament must now approve the directive before member countries turn it into law.) A number of European countries already have mandatory take-back programs. Depending on the country, costs are borne by consumers, municipalities, or manufacturers.[25]

Bridging production and disposal through take-back programs may spur manufacturers to design products for easier disposal. Currently, one obstacle to disposal is the high toxic load of many products. A computer monitor, for example, contains 1.8–3.6 kilograms of lead, a heavy metal that damages the nervous system and poisons blood cell development.[26] Monitors already account for nearly 40 percent of the lead in U.S. landfills.[27] Cadmium, found in computer batteries, is recognized to increase the risk of cancer, damage the developing fetus, and harm the reproductive system.[28] Flat panel screens contain mercury, which can form organic compounds that damage the developing nervous system.[29] The logic is straightforward: putting fewer toxic chemicals into electronic products will mean less hazardous waste to throw away later.

# Toxic Waste Largely Unseen

*Anne Platt McGinn*

Some 300–500 million tons of hazardous waste were generated worldwide each year during the past decade.[1] This amounts to roughly 50–83 kilograms per person in 1999 alone—and there is no end in sight.[2]

Under the 1989 Basel Convention on the Control of Transboundary Movements of Hazardous Wastes and Their Disposal, wastes are classified as hazardous if they exhibit one or more hazardous characteristics and appear on a list of waste streams or if they contain specified hazardous constituents, such as asbestos, heavy metals, and several other chemicals.[3] (See Table 1.) Many industries create hazardous waste, including medical care, mining, petrochemicals, and pesticides and plastics manufacturing.[4]

*Links:* pp. 82, 84, 110, 114

Industrial countries create more than 80 percent of the world's hazardous waste.[5] The United States is the largest producer, accounting for an estimated 260 million tons in 1997, including heavy metals, solvents, and toxic sludge.[6] By comparison, 39 countries that have ratified the Basel Convention reported generating 252 million tons of hazardous and other possibly dangerous wastes in 1998.[7] Russia and Uzbekistan accounted for half of this total.

Of the few countries that have filed more recent data with the Basel Convention secretariat, several reported increases in 1999.[8] China claimed a 2.6-percent increase in hazardous waste generation from 1998 to 1999, while the United Kingdom posted 20-percent growth during this time.[9]

Yet these self-reported data are an incomplete measure of the problem because fewer than one third of the 149 countries that have ratified the Basel Convention actually filed a national report with the secretariat in 1998, and many countries admit the data in these reports are unreliable.[10]

While hazardous waste generation continues with no signs of slowing, the global waste equation has grown more complex in response to the Basel Convention. The treaty aims to reduce cross-border movements of hazardous wastes while minimizing their generation, to promote disposal close to site of origin, and to prohibit trade with countries that lack the capacity to manage wastes in an environmentally sound manner.[11] In 1995, a group of developing countries and the European Union passed an amendment to the convention to prohibit the export of wastes from industrial to nonindustrial countries.[12]

This amendment is not yet legally binding (35 more ratifications are needed for it to enter into force), but most countries abide by its prohibition voluntarily.[13] The United States is a notable exception: U.S. officials have argued that the Basel ban may prevent some legitimate recycling activities and could inhibit trade.[14] (The United States signed the Basel Convention itself in 1989 but has not yet ratified it.)[15]

In addition to the global ban on exports, many countries have passed national laws and acceded to regional agreements to prohibit imports of hazardous wastes.[16] Regional bans in Africa and Latin America, for example, now forbid importing asbestos, unregistered pesticides, and other hazardous products.[17] As a result of these legal agreements, actual and attempted waste transfers between industrial and developing countries have declined significantly in recent years.[18]

Today, about 10 percent of all hazardous waste is moved across an international border, mostly among industrial nations.[19] The primary exporters are Australia, Germany, the Netherlands, the United Kingdom, and the United States.[20]

Canada has recently become a dumping ground for toxic waste in North America, owing to its less restrictive regulations.[21] Between 1993 and 1999, imports of hazardous waste to Canada from the United States and Mexico jumped 400 percent.[22] In fact, in 1999 Canada accepted more than twice as much hazardous waste from the United States as Mexico did.[23]

While hazardous waste transfers among rich nations continue largely unrestrained, waste shipments between developing countries are a growing concern.[24] Illegal trade is also ongoing and difficult to stop.[25]

The pressures that contributed to all this

## Table 1: Hazardous and Nonhazardous Waste in the Basel Convention

| List | Examples |
| --- | --- |
| Annex VIII of the Basel Convention— characterized as hazardous | Metal wastes and waste consisting of alloys of arsenic, cadmium, lead, and mercury<br>Waste lead-acid batteries, whole or crushed<br>Waste electrical and electronic assemblies or scrap<br>Waste asbestos |
| Annex IX of the Basel Convention— not characterized as hazardous unless they contain hazardous materials | Iron and steel scrap<br>Metal-bearing wastes arising from the melting, smelting, and refining of metals<br>Other ceramic, solid plastic, paper, rubber, and textile wastes |
| Working list of wastes awaiting classification | Polyvinyl chloride (PVC) waste<br>PVC-coated cables<br>Residues from industrial waste disposal operations |

Source: UNEP, Report of the Fourth Meeting of the Conference of the Parties to the Basel Convention (Geneva: 18 March 1998).

trade in the first place—increasing volumes of waste, higher disposal costs in industrial countries, and different national waste control standards—have increased since the mid-1990s.[26] Without efforts to reduce the overall quantities of waste, progress achieved during the first decade of the Basel Convention could be undermined quickly.

Because the treaty controls waste intended for final disposal but not for recovery or recycling, countries now prefer to label waste shipments for recycling.[27] In 1998, an estimated 11 percent of exported wastes was burned, landfilled, or otherwise disposed of—while the other 89 percent was recycled.[28] While this sounds like a preferable environmental option, many recycling and recovery operations are seen as a pretext for sending hazardous materials to countries for use in energy production, road building, construction, fertilizer manufacturing, and substandard and hazardous recycling operations.[29] Such uses expose greater numbers of people to health risks and spread the contamination.

Another form of toxic transfer is the relocation of industries and technologies that generate hazardous materials from industrial to developing countries.[30] For example, the global shipbreaking industry has recently shifted its focus from industrial countries to Asia.[31] Shipbreaking involves dismantling vessels contaminated with explosive gases, asbestos, PCBs, and other toxins. Most of the world's shipbreaking is now done by migrant workers in Asia, with little or no health protections.[32]

People who live near toxic waste dumps have reported increased vulnerability to certain cancers, birth defects, and low birth weight.[33] Babies whose mothers lived within three kilometers of a landfill were found to have a 33-percent higher risk of congenital birth defects than babies living three to seven kilometers away, based on European data.[34] One study concluded that living near a hazardous landfill poses the same risk of having a baby with low birth weight as smoking during pregnancy.[35]

Despite the growing number of countries that have ratified the Basel Convention and its amendment and the global crackdown on trade between rich and poor countries, new hazardous waste continues to be produced at the rate of about a million tons per day and is transferred in many forms, largely unmonitored.[36] Only by incorporating cleaner technologies and safer products can societies prevent the creation, use, and proliferation of hazardous materials and address the underlying causes of the ever-growing waste crisis.

# Rio Treaties Post Some Success

*Jessica Dodson*

The Rio Earth Summit in 1992 sharpened the world's focus on environmental issues, bringing them to the front of global consciousness to an unprecedented degree. The World Summit on Sustainable Development in Johannesburg in September 2002 will surely remind the world of the promises made in Rio and the successes it can claim. Since 1992, six major multilateral treaties have been adopted, while older agreements have been strengthened and refined by the addition of protocols and amendments.[1] (See Table 1.)

The six multilateral treaties, the so-called Rio Conventions, represent important advances in international environmental law.[2] Despite a broad global consensus that environmental issues are a priority, many countries are reluctant to sign such treaties, particularly those requiring substantive changes in social or economic behavior.[3] As result, in the post-Rio era, treaty drafters have increasingly used incentives in order to entice countries to take part.[4]

Links: pp. 112, 126

Since Rio, industrial nations have generally acknowledged a special responsibility for environmental degradation, and to secure the participation of developing countries in environmental treaties, they have agreed to finance part of any implementation costs. But industrial countries also often need incentives to join treaties. For instance, both Japan and Russia were offered a range of concessions in order to elicit their approval of a refined version of the Kyoto Protocol at Marrakech in November 2001.[5]

Along with incentives, environmental treaties now regularly include sanctions and penalties in order to enforce compliance. Trade sanctions, used against states not abiding by their commitments in regimes such as the Montreal Protocol on the ozone layer and the Convention on International Trade in Endangered Species (CITES), remain an important enforcement tool. Such measures are often a source of contention, as states balk at being held accountable for violations.[6]

There is also a potential for conflict between global free trade rules and the growing body of environmental law. World Trade Organization (WTO) trade rules forbid restriction of the free circulation of goods, including goods whose production aggravates environmental damage. As treaties continue to include trade-restriction clauses, the risk of violating WTO free trade rules grows.[7]

While environmental diplomacy has unquestionably grown more sophisticated and prominent in the decade since Rio, have the new treaties stemmed the tide of environmental deterioration? In some cases, the compromises made during treaty negotiations may render treaty provisions too weak to address problems adequately—a charge frequently levied against the Kyoto Protocol.[8]

In assessing outcomes, it should be noted that the more specific the obligation, the easier it is to actually judge compliance and measure the treaty's impact. Thus adherence to the Montreal Protocol, CITES, or the treaty on persistent organic pollutants is much easier to measure than compliance with treaties on biodiversity or desertification, where obligations are broader and means of implementation not specified.[9] The Montreal Protocol has been lauded as particularly effective in reducing the incidence of ozone-depleting substances, whereas phenomena such as the loss of biodiversity and the trade in endangered wildlife have continued or accelerated despite treaties intended to reverse these trends.[10]

Although results clearly vary from treaty to treaty, most have had at least some positive effect on the problem they address.[11] There remains ample room for improvement, however. The proliferation of environmental agreements represents an unquestionable stride forward, but it may also provide a false sense of security that enough is being done. Treaty effectiveness must be assessed through the systematic collection of data and information. This will help analyze weaknesses in the environmental regime, design more effective treaties in the future, and pave the way for wider participation by eliminating the reticence fostered by uncertainty and denial.

## Table 1: The Rio Conventions

### Convention on Biological Diversity, 1992 (182 parties)

Provides broad guidelines for the conservation of biodiversity at the national level and requires countries to formulate national biodiversity strategies and file national reports. Recognizes national sovereignty over biological resources and principle of Prior Informed Consent before resources may be transferred out of a country, stipulating that biodiversity use must be sustainable and resulting benefits must be equitably shared between the source and receiving countries. Subsequent Biosafety Protocol in 2000 (11 parties, requires 39 more to enter into force) provides strong enunciation of the precautionary principle, allowing states to decline to import products that "may contain" genetically modified organisms.

### U.N. Framework Convention on Climate Change, 1992 (186 parties)

Richest and most industrialized countries agree to adopt policies to stabilize greenhouse gas emissions at 1990 levels by 2000. Treaty introduces innovative procedures for implementation such as an emissions trading system, the Clean Development Mechanism, and Joint Implementation of commitments. Subsequent Kyoto Protocol in 1997 (47 parties, requires at least 8 more; will enter into force when states representing 55 percent of 1990 carbon dioxide emissions have ratified) specifies 5.2-percent reduction in overall emissions from 1990 levels by 2012 and delineates other specifics of implementation procedures.

### Convention to Combat Desertification, 1994 (178 parties)

Designed to facilitate regional efforts to counter desertification; creates a network of four regions—Africa, Asia, Latin America and Caribbean, and Northern Mediterranean. Each area can design and implement a plan tailored to local needs, but funding is primarily the responsibility of the afflicted states, with supplementary assistance from the donor community.

### UN Agreement Relating to the Conservation and Management of Straddling Fish Stocks and Highly Migratory Fish Stocks, 1995 (31 parties)

Entered into force in late 2001; advocates a cooperative, precautionary approach to management and conservation of relevant fish stocks. Coastal states and those fishing in international waters must adopt national measures to restore stocks to levels capable of producing maximum sustainable yields. Includes provisions allowing parties to board and inspect vessels of other parties on the high seas in order to verify compliance. Also encourages regional planning and information exchange, recognizes the needs of developing states and subsistence fishers, and contains provisions on pollution control, related ecosystems, and domestic monitoring and compliance.

### Rotterdam Convention on the Prior Informed Consent Procedure for Certain Hazardous Chemicals and Pesticides in International Trade, 1998 (18 parties, requires 32 more to enter into force)

Building on nonbinding procedures developed over 10 years, exporting states must receive explicit permission from importing states before shipments of 27 types of restricted substances may take place. Safety and labeling requirements specified for the handling of these substances. States refusing shipments containing a chemical must halt domestic production of the substance, avoiding conflict with trade rules.

### Stockholm Treaty on Persistent Organic Pollutants, 2000 (5 parties, requires 45 more to enter into force)

Regulates the production and use of 12 persistent, toxic substances. The 9 Annex A chemicals are slated for elimination, while Annex B lists chemicals such as DDT that are subject to restricted use. Also mandates the identification and elimination of stockpiles, products, and wastes containing persistent organic pollutants.

*Source*: See endnote 1.

# Economy and Finance Features

© DIGITAL VISION

OZONE FRIENDLY

CONTAINS **NO** CHLOROFLUORO C A R B O N PROPELLANT ALLEGED TO DAMAGE THE **O**ZONE LAYER

Foreign Aid Spending Falls
Charitable Giving Widespread
Cruise Industry Buoyant
Ecolabeling Gains Ground
Pesticide Sales Remain Strong

# Foreign Aid Spending Falls

*Hilary French*

Ten years ago, the Rio Earth Summit attempted to bridge the interests of countries of the North and the South in forging a sustainable development path through what is sometimes called the Rio bargain. The essence of this deal was that industrial and developing countries would agree to implement the range of environmental provisions contained in Agenda 21 and other Rio documents, but that industrial countries would provide substantial financial resources to help others accomplish this.[1] As the World Summit on Sustainable Development in Johannesburg approaches, frustration is running high in many quarters over a failure of industrial countries to uphold their end of this bargain.

Agenda 21, the lengthy action plan that emerged from the Rio conference, estimated that $125 billion in foreign aid would be needed to put the plan into practice, on top of substantially stepped-up spending by national governments.[2] This sum was widely viewed as unrealistic at the time, as it amounted to twice the overall spending on foreign aid.[3] But northern governments nonetheless agreed to strive to meet it, in part by reaffirming the commitments of many donor countries to contribute 0.7 percent of their gross national product (GNP) annually to development assistance.[4]

*Links: pp. 62, 90, 148*

But in the decade since Rio, aid spending has declined substantially rather than increased. According to Organisation for Economic Co-operation and Development figures, official development assistance amounted to $54 billion in 2000, down from $73 billion in 1992 (in 2000 dollars). (See Table 1.) Aid spending as a share of donor nations' GNPs also declined, from 0.33 percent to 0.22 percent.

Spending levels vary greatly by individual donor country. In relative terms, Denmark leads the list, consistently contributing more than 1 percent of its GNP in aid, with the Netherlands, Sweden, and Norway following close behind. The United States ranks as the least generous donor by this measure, spending just 0.1 percent of its national income.

Several conditions are thought to have contributed to the decline in aid spending over the last decade, including the end of the cold war, large fiscal deficits in donor countries during the early to mid-1990s, and the growth of private capital flows into many parts of the developing world. An additional factor has been growing skepticism from many quarters about the effectiveness of development aid in combating poverty and addressing other critical social and environmental challenges, particularly in countries beset by corruption.[5]

Despite the shortcomings of many foreign aid programs, there can be little doubt that the overall decline in aid spending over the last decade has made it more difficult to fund key environmental and social programs adequately. The Global Environment Facility (GEF) is a case in point. GEF's mandate is to finance the additional costs that developing countries incur in responding to global environmental problems, including climate change, ozone depletion, the loss of biological diversity, the degradation of international waterways, and the spread of persistent organic pollutants.[6]

Projects financed by the GEF have, among other things, helped Ethiopian farmers learn new ways to preserve genetic diversity in local agriculture; encouraged a partnership between an environment group, a local government, and a cement plant to preserve the Dana Nature Reserve in Jordan; and helped thousands of households, health clinics, and schools in some 20 countries to install solar power systems.[7] Over the last decade, the GEF has committed $3.4 billion in grants to over 650 projects in 150 countries, an average of some $300 million per year.[8] But raising even this relatively small sum from donor governments has proved to be a continuing challenge.

Like the GEF, other environmental institutions have also suffered from scarce funding. Budgets of the small offices charged with administering critical environmental treaties such as the Montreal Protocol and the biological diversity convention generally range from $1–10 million, and UNEP has struggled to maintain its annual budget of roughly $100 million.[9] (In comparison, the U.S. Environmental Protection Agency had a budget of $7.8 billion in 2001, while the U.S. military budget was over $311 billion and global military expenditures in 2000 added up to more than $780 billion.)[10]

Programs aimed at reducing poverty and other pressing social problems are also starved for cash. For instance, in 2001 U.N. Secretary-General Kofi Annan called on donor nations to contribute $7–10 billion a year to a global fund to finance prevention and treatment programs for AIDS, tuberculosis, and malaria, three of the world's major killers.[11] Nine months later, the fund had attracted only $2 billion in pledges, and less than half of that is expected to be delivered in 2002.[12]

In September 2000, world leaders gathered in New York for the U.N. Millennium Summit, where they adopted a set of aggressive social goals for 2015, including halving the share of the world's people living in extreme poverty, suffering from hunger, and lacking access to clean drinking water; reducing maternal mor-

tality by three quarters; and cutting child mortality by two thirds.[13]

Some $50 billion in additional aid spending will be needed to meet these targets, according to a report prepared for Secretary-General Annan as an input into the March 2002 International Conference on Financing for Development in Monterrey, Mexico.[14] If all donor countries were to meet the 0.7-percent goal, an additional $100 billion in annual spending could be raised—more than enough to cover these costs.[15]

Some governments and activists are pushing for donors to make a renewed commitment to the 0.7-percent aid target. The United Kingdom has been vocal on this score, with Chancellor of the Exchequer Gordon Brown calling in December 2001 for a "new Marshall Plan" to fight poverty and other social ills that threaten both human security and international stability.[16]

## Table 1: Development Assistance Contributions, Top 15 Countries and Total, 1992 and 2000

| Country | 1992 | | 2000 | |
|---|---|---|---|---|
| | Total (million 2000 dollars) | As Share of GNP (percent) | Total (million 2000 dollars) | As Share of GNP (percent) |
| Denmark | 1,621 | 1.02 | 1,664 | 1.06 |
| Netherlands | 3,207 | 0.86 | 3,135 | 0.84 |
| Sweden | 2,865 | 1.03 | 1,799 | 0.80 |
| Norway | 1,483 | 1.16 | 1,264 | 0.80 |
| Belgium | 1,014 | 0.39 | 820 | 0.36 |
| Switzerland | 1,327 | 0.46 | 890 | 0.34 |
| France | 9,634 | 0.63 | 4,105 | 0.32 |
| United Kingdom | 3,778 | 0.31 | 4,501 | 0.32 |
| Japan | 12,990 | 0.30 | 13,508 | 0.28 |
| Germany | 8,834 | 0.39 | 5,030 | 0.27 |
| Australia | 1,182 | 0.35 | 987 | 0.27 |
| Canada | 2,930 | 0.46 | 1,744 | 0.25 |
| Spain | 1,769 | 0.26 | 1,195 | 0.22 |
| Italy | 4,802 | 0.34 | 1,376 | 0.13 |
| United States | 13,640 | 0.20 | 9,955 | 0.10 |
| All Countries | 73,055 | 0.33 | 53,737 | 0.22 |

Source: Organisation for Economic Co-operation and Development (OECD), "ODA Steady in 2000; Other Flows Decline," 12 December 2001; OECD, Development Assistance Committee, *Development Assistance Committee Online*, updated 30 January 2002; OECD, Development Assistance Committee, *Development Co-operation 1993* (Paris: 1994), pp. 168–69.

# Charitable Giving Widespread

*Brian Halweil*

The vast majority of people in the industrial world give money to charity, amounting to billions of dollars each year, although per capita giving varies widely among nations.[1] (See Table 1.)

Because of different tax laws and accounting methods, national statistics on charitable giving are not always comparable. Statistics are most readily available for giving in the industrial nations of North America, Europe, and Asia, due to their greater wealth and better accounting. But charitable giving seems to be a universal phenomenon. In particular, informal modes of giving—through family ties, churches, and clothing and food donations, for example—are widespread in both rich and poor nations.[2]

Links:
pp. 118, 148

Charitable giving is a proxy for how much people are concerned about community affairs or those less fortunate, but giving levels may also indicate social needs that are not being met otherwise. For instance, since most charitable giving is for domestic causes, some researchers have suggested that the relatively low levels in Europe compared with the United States result from the higher tax levels and stronger social welfare policy in Europe.[3]

Levels of giving are also affected by the economic situation, unemployment levels, and tax laws.[4] Giving usually follows the movement of the economy—with more donations in boom times, and less in recessions.[5] The enormous economic expansion in the United States between 1996 and 2000 coincided with a steep rise in giving, particularly among very wealthy individuals.[6]

Historical trends in charitable giving are available for only a few nations, but generally show that while total giving has increased in real terms, giving per person has increased only modestly or declined.[7] Canadians donated more than $5 billion to charities and nonprofits in 2000, for example, an increase of 11 percent since 1997; over the same period, donations by the average person rose 8 percent.[8]

Over the last 30 years, giving in the United States has grown at an average rate of 2.6 percent (adjusted for inflation), more than doubling from $93 billion in 1970 to $203 billion in 2000.[9] Individual giving accounts for over 80 percent of all donations in the United States, with foundation and corporate donations providing 12 percent and 5 percent respectively.[10] Since 1970, however, giving by foundations and corporations has grown more rapidly than individual donations. The figure for foundations, adjusted for inflation, increased from $8.4 billion in 1970 to $24.5 billion in 2000, while corporate giving grew from $3.6 billion to $10.9 billion.[11]

Foundation gifts are particularly dependent on the state of the stock market, since in the United States these organizations are required by law to give out a certain percentage of their total holdings. This means that foundation giving has increased in recent years, but contracted severely with the stock market correction toward the end of 2001.

Many nations do not have the tax structure to support giving or the organizations for people to donate to. Before 1996, for instance, Canadians could claim tax credits for donations of up to 20 percent of their taxable income. In 1996, the bar was raised to 50 percent of taxable income—and giving jumped 14 percent over the previous year.[12] In Japan, where corporations enjoy considerably greater latitude and tax deductions from donations than individuals do, corporations contribute

## Table 1: Individual Giving in Selected Industrial Nations, Late 1990s and 2000

| Nation | Annual Giving Per Person (dollars) | Share of Population (percent) |
|---|---|---|
| Canada | 259 | 91 |
| United States | 953 | 89 |
| Netherlands | 275 | 76 |
| United Kingdom | 180 | 68 |
| Japan | 15 | n.a. |
| France | 380 | n.a. |

*Source*: See endnote 1.

$3.7 billion each year compared with just $205 million by individuals.[13] (The latter figure is thought to be severely underestimated, however, because individual donations are not tax-deductible and are therefore not reported.)[14]

In the United States, individuals tend to give more, as a share of their total wealth, than corporations do. Individual donations in 2000 amounted to 1.8 percent of income, whereas corporate donations amounted to 1.2 percent of pretax income in the same year.[15]

In any given country, a relatively small group of people tend to account for the vast majority of donations. In Canada, for instance, the top 25 percent of donors were responsible for 82 percent of total donations in 2000.[16] Although the wealthy tend to give more in absolute terms, they generally give less as a percentage of their income. For example, Canadians with an annual income of less than $20,000 give nearly three times as high a share of their income as those with incomes of $100,000 or more.[17]

The likelihood of making donations increases with age, income, education level, religious involvement, and marriage.[18] In both the United States and Canada, people with a religious affiliation gave twice as much, on average, as donors without one.[19] And one survey indicated that college graduates donated 50 percent more of their income than high school graduates.[20] Surveys from several nations also indicate that women give more often and more generously than men.[21]

Although people in both Europe and North America give heavily to religion, North Americans particularly favor such causes. In Canada, more than half of all donations are made to religious organizations; health organizations and social service organizations captured 20 and 10 percent of the funds respectively.[22] In the United States, over one third of all giving went to religion, including half of the money donated by individuals; education and health were a distant second and third.[23]

In contrast, the Dutch give roughly 15 percent of total donations to each of the following: health organizations, sports and recreational

organizations, social welfare groups, and international aid.[24] Medical causes receive almost one quarter of all British donations, while children and young people receive some 16 percent.[25]

Although most donations never cross a border—that is, most money is donated for local or national causes—international causes are increasingly popular, particularly in Europe. In the United Kingdom, international aid accounted for 9 percent of all donations, compared with just over 1 percent in the United States.[26]

Surveys indicate that people generally give because they feel compassion toward those in need, because they owe something to their community, or because they have been personally affected by a cause.[27] Sudden disasters—from earthquakes to plane crashes—can also prompt charitable giving. Sixteen percent of Americans who gave to charities associated with the September 11th attacks on New York and Washington had not given to any charitable cause in the previous year.[28]

People give in different ways, including donations of clothing or food or by volunteering their time. In 2000, 6.5 million Canadians—27 percent of the population aged 15 and older—volunteered for charities and nonprofits, an average of 162 hours per person.[29] The formal volunteer work force in the United States in 2001 represented the equivalent of over 9 million full-time employees, at a value of $239 billion.[30]

And the more money that an individual gives, the more likely he or she is to provide other types of support, including volunteering, giving directly to other people, and participating in community organizations.[31] In Canada, among the top 25 percent of donors in terms of total value given, almost half also volunteered time and nearly three quarters were members of an organization or group.[32] In contrast, of Canadians who did not give donations, just 11 percent volunteered and 28 percent belonged to organizations.[33]

# Cruise Industry Buoyant

*Lisa Mastny*

The number of people taking a cruise vacation more than doubled between 1990 and 2000, to 9.8 million passengers annually, according to U.K.-based analyst G.P. Wild Ltd.[1] (See Figure 1.) The global cruise industry has grown on average 8 percent a year over the past decade, nearly twice as fast as tourism overall.[2] Demand is expected to again double by 2010, to an estimated 20.7 million passengers annually.[3]

Cruises were once a luxury for upscale travelers: as recently as 1970, only about half a million people took one.[4] But today many cruise lines offer inexpensive promotional packages as well as a wide range of on-board amenities to attract more mainstream tourists.[5] Larger vessels resemble "floating cities," carrying more than 3,000 passengers and 1,000 crew, and boasting spas, conference centers, and even skating rinks.[6] In 2001, the world's cruise fleet totaled some 163 ships, with an additional 42 new vessels slated for construction by 2005.[7]

The world's four major cruise companies—Carnival Corporation, Royal Caribbean Cruises Ltd., P & O Princess Cruises, and Star Cruises—controlled roughly three quarters of the market in the late 1990s, earning combined revenues of $10.4 billion in 2000.[8] Industry concentration continued in late 2001, when Royal Caribbean and P & O Princess—which together carried more than 3 million passengers on 41 ships in 2000—announced a $6-billion merger to create the world's largest cruise group.[9]

Nearly 70 percent of cruise passengers come from North America, while 21 percent come from Europe and the rest mostly from Asia and the Pacific.[10] The Caribbean remains the top destination, accounting for 45 percent of global capacity, followed by the Mediterranean (13 percent), Europe (8 percent), Alaska (8 percent), and the U.S. West Coast (3 percent).[11] Since 1991, the number of passengers visiting Alaska has jumped by 10 percent a year, to 630,000 annually; cruises now account for about 80 percent of that state's tourism business.[12] All told, cruise-related activities contributed $18 billion to the U.S. economy in 2000.[13]

But many cruise destinations, particularly in the developing world, do not see widespread benefits from vessel visits. Port countries typically earn money from docking fees (which are often low), taxes, sales of fuel and fresh water, services such as waste disposal, and whatever passengers spend on shore.[14] Yet with the rapid expansion of on-board offerings, many cruise passengers spend relatively little time or money at their ports-of-call.[15] Meanwhile, most on-board restaurants and shops import food and supplies from the United States or Europe rather than buying locally.[16]

Nearly all major cruise ship owners sail their vessels under foreign flags, taking advantage of lower corporate taxes and wages in many countries.[17] Carnival Corporation, incorporated in Panama, paid just $19 million in taxes on $2 billion in operating income in the mid-1990s, while Royal Caribbean saves an estimated $30 million a year by registering its ships in Liberia and Norway.[18]

Vessel re-flagging also makes it easier for ship owners to flout international labor, safety, or other standards by allowing them to register ships in countries with weaker laws or enforcement.[19] The world's top cruise lines recruit as many as 90 percent of their employees from the

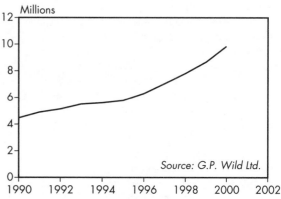

Figure 1: Global Demand for Cruises, 1990–2000

international work force—a single crew can have staff from as many as 60 nations.[20] Some of these workers face 14-hour days and wages below $2 an hour, with little or no job security.[21]

Ship operators are also notorious for lax adherence to laws regulating the disposal of sewage and other wastes.[22] The San Francisco–based Bluewater Network reports that on a one-week voyage a typical cruise ship generates some 795,000 liters of sewage; 3.8 million liters of graywater (from sinks, showers, and laundry); 95,000 liters of oily bilge water; eight tons of garbage; 416 liters of photo chemicals; and 19 liters of dry-cleaning waste.[23] Vessel smokestacks can emit high levels of air pollutants, including nitrous oxides, sulfur dioxide, and carbon dioxide.[24] Overall, the world's cruise ships discharge some 90,000 tons of raw sewage and garbage into the oceans each day.[25]

International maritime law permits vessels to release specified levels of pollutants overboard, provided the waste is treated or diluted or is discharged a certain distance from shore.[26] Yet many cruise ships dump their wastes illegally—and most of this goes undetected. In 1999, Royal Caribbean was fined a record $18 million for releasing excess oily bilge and other pollutants into U.S. waters and for attempting to cover up its crime.[27] And a 2000 study of cruise ship effluents in Alaskan coastal waters found that 57 percent of sewage samples failed to meet U.S. federal standards for fecal coliform bacteria, while 68 percent failed to meet standards for suspended solids.[28]

Cruise ships have other environmental impacts. To accommodate larger vessels, many countries dredge deep-water harbors or modify their coastlines, destroying coastal ecosystems in the process. And when ships dock, their massive anchors and chains can break coral heads and devastate underwater habitats: in 1994, a scientist in the Cayman Islands reported that more than 120 hectares of reefs had been lost as a result of cruise ships anchoring in George Town harbor.[29]

In an effort to clean up their acts, many cruise companies are trying to "green" their management and operations. Simple steps include recycling plasticware and using recyclable and reusable containers.[30] Some lines, like Holland America, are outfitting new vessels with on-board sewage treatment plants, incinerators, or co-generation incinerators that harness energy from waste burning.[31]

In 2001, the International Council of Cruise Lines—a group that represents the top 16 cruise lines and whose members' 100 ships carry more than 7 million passengers annually—adopted new mandatory waste management standards.[32] Companies risk losing membership if they violate the guidelines, which include rules for disposal of wastewater, batteries, and toxic chemicals and which call for better compliance with national and international environmental laws.[33]

Many smaller cruise operators are embracing voluntary "codes of conduct" to regulate their impacts or are participating in schemes that certify good environmental practice.[34] The 46 members of Antarctica's tour operators' association now follow a strict code that includes landing no more than 100 people per site at a time and making sure visitors do not disturb wildlife.[35] And the new SmartVoyager program in the Galapagos has so far certified 5 of the area's more than 80 cruise operations for voluntarily meeting benchmarks set for maintenance and operations, docking, and fuel and wastewater management.[36]

Governments are also taking action. New legislation in Alaska, for instance, regulates graywater and airborne emissions from larger cruise vessels, allows inspectors to fine violators, and charges $1 per passenger to fund state pollution control.[37] But the rules exempt certain hazardous wastes, and critics worry that the continued rapid rise in passenger numbers could outweigh regulatory efforts.[38]

# Ecolabeling Gains Ground

*Lisa Mastny*

As the demand for environmentally friendly products grows, manufacturers, governments, and nongovernmental groups have expressed rising interest in "ecolabeling."[1] Ecolabels are seals or logos used to indicate that a product has met a specified set of environmental or social standards.[2]

Although ecolabeling schemes vary widely, they typically reward a product for its environmental soundness during one or more stages of its life cycle, including production, packaging, use, or disposal.[3] Some programs focus on a single product: the Mexico-based Forest Stewardship Council, for instance, grants its seal to wood products that have met certain social and environmental standards during harvesting, manufacturing, and distribution.[4] In contrast, the U.S.-based Green Seal program evaluates and certifies a wide range of products, including paints, engine oil, and air conditioners.[5] Worldwide, ecolabels can now be found on everything from organic foods to tourism destinations.[6] (See Table 1.)

*Link*: p. 132

Ecolabeling programs exist at the national, regional, and global levels. The first national scheme, Germany's Blue Angel, was launched in 1978 and now awards its seal to some 3,900 products and services—from batteries to car washes.[7] Subsequent programs include India's Ecomark and Singapore's Green Label, both developed in the early 1990s.[8] Currently, at least 24 countries have national ecolabeling programs, and many more are developing them.[9] At the regional level, schemes include the Nordic Swan, which certifies more than 3,000 different products in Europe's Nordic countries, and the European Union's Flower Eco-label, which has been applied to 400 products.[10]

Ecolabeling schemes serve a dual purpose. They can help encourage the design, production, marketing, and use of more environmentally sound products and services.[11] But they also provide consumers with valuable information about the range of preferable products, helping them to make more informed purchasing choices.[12] A 1996 Green Gauge poll found that 45 percent of Americans had bought specific products because the labels stated they were environmentally safe or biodegradable.[13]

The most effective ecolabeling programs have been developed with input from consumers as well as industry and environmental groups.[14] Independent certification bodies evaluate whether a product conforms to a set of meaningful and consistent standards for environmental protection or social justice.[15] Certifiers can include government agencies (such as the U.S. Department of Agriculture), nongovernmental groups (the Rainforest Alliance), professional or private groups (Green Seal), or international accreditation bodies (the Marine Stewardship Council).[16]

Ecolabeling works best when the labels rely on a set of clearly defined and verifiable standards—such that a single label means the same thing if used on a wide range of products.[17] For many product areas, however, several competing ecolabels now exist, creating the potential for consumer confusion. For instance, at least three different bodies worldwide independently certify sustainably harvested wood, and more than 100 schemes reward environmentally or socially responsible tourism.[18] One way to resolve this problem is to develop a universal labeling standard for a specific industry or product, though this is generally a challenge.[19]

Ecolabeling faces other obstacles. Critics worry that some schemes rely on a relatively low standard, in order to reach out to more manufacturers and to spur greater interest in producing or buying environmentally preferable goods.[20] For instance, the U.S. government's Energy Star label, which rewards energy-efficient appliances and other products, is so inclusive that in 1995 an estimated 85–90 percent of computers qualified for it.[21] But the program's inclusiveness may prevent it from spurring the development of more cutting-edge energy-saving technologies.[22]

Ecolabeling faces economic challenges as well. Many certification schemes charge a fee for evaluation, which may be too high for smaller companies or producers and can limit expansion of the market.[23] Companies may also pass the costs of certification on to consumers, boosting the prices of ecolabeled products.[24] And there is concern that programs that rely on self-certifying may simply allow companies to "buy" their way to a green label.[25]

Consumers also need to distinguish genuine ecolabels from more general claims manufacturers make about the environmental soundness of their products. Many of these claims—such as "dolphin-safe," "antibiotic-free," "biodegradable," or "elemental chlorine-free"—may be accurate, but they are not always independently verified.[26] At times, the labels can be highly ambiguous and may only confuse consumers, as with products that claim to be "environmentally friendly" or "Earth smart."[27]

Moreover, just because a product carries an ecolabel does not mean it is necessarily the most environmentally sound option. In some cases, reusing or doing without may be environmentally preferable to buying a labeled product (reusing a cloth towel instead of buying recycled paper towels, for instance).[28] And some products may be awarded ecolabels even though the overall environmental track record of the manufacturer is poor.[29] Ecolabels can also be relatively narrow in scope, focusing only on one specific attribute: a label may reward a product for its energy efficiency, but hide the fact that it also contains toxic materials.[30]

Ultimately, the success of ecolabeling will depend on whether trusted, reliable standards can be set and on the degree to which the industry and consumers embrace it worldwide.

## Table 1: Ecolabeling Schemes for Selected Products

| Product | Example | Description |
| --- | --- | --- |
| Forest products | Forest Stewardship Council | Grants its logo to products obtained from forestry operations that meet specified standards for sustainable management and harvesting. Has certified more than 25 million hectares of forests in 54 countries. |
| Agriculture | Rainforest Alliance's Conservation Agriculture Network | Awards its logo to coffee, banana, cocoa, and citrus farms that adhere to specified environmental and social standards. As of June 2001, had certified 51,600 hectares on 218 farms or cooperatives in nine countries, mainly in Central America. |
| Seafood | Marine Stewardship Council | Awards its logo to fisheries that meet a set standard for environmentally responsible management and practice. Has certified six fisheries so far for rock lobster, cockles, hoki, mackerel, herring, and salmon. |
| Coffee | Smithsonian Migratory Bird Center | Awards its Bird Friendly® seal of approval to coffees from Latin America that have been independently certified to meet specified standards for shade farming and organic production. |
| Energy | Green-e-Renewable Electricity Certification Program | Rewards electricity services that obtain at least half their supply from renewable sources; the first voluntary certification and verification program for green electricity in the United States. |
| Tourism destinations | European Blue Flag Campaign | Awards a yearly label to some 2,750 beaches and marinas in 21 European countries for their high environmental standards and sanitary and safe facilities. Credited with improving the quality and desirability of European coastal sites. Also being adopted in South Africa and the Caribbean. |

*Source*: See endnote 6.

# Pesticide Sales Remain Strong

*Brian Halweil*

Although down slightly in recent years, global pesticide sales have increased 15-fold since 1950, from $2.8 billion (2000 dollars) in 1950 to $42 billion in 1999.[1] (See Figure 1.) Sales for agricultural use increased from $1.8 billion to just over $30 billion in the last half-century, while sales for industrial applications—including home and building pest control, road and highway use, and golf course maintenance—have grown from just under $1 billion to $12 billion.[2]

The growth of agricultural pesticide sales has slowed in recent decades, declining from an average of 8–10 percent per year in the 1960s and 1970s to under 2 percent in the 1980s and virtually no growth during the 1990s.[3] Agricultural markets are nearly saturated in the industrial world, which accounts for 65 percent of global sales.[4] Industrial uses of pesticides, however, which now account for more than one quarter of total pesticide use, continue to expand at more than 3 percent each year.[5] By 2010, industrial sales are projected to represent 30 percent of the total market.[6]

*Links*: pp. 26, 112, 114, 130

The United States is about 40 percent of the world market for household pesticides, with annual sales exceeding $1 billion.[7] China is the second largest market for this category, with $580 million in sales.[8] Americans also lead the way in garden pesticide purchases, with annual sales of $1.5 billion, followed by the United Kingdom at $155 million.[9] Of the $850-million-dollar market for "turf" pesticides, roughly half is used on the world's golf courses and most of the remainder gets applied to American lawns.[10]

North America, Europe, and Japan account for 65 percent of global agricultural pesticide sales. The United States, Japan, and France alone account for over 40 percent of global sales.[11] Latin America buys about 13 percent of these products, while Asia (excluding Japan) purchases about 12 percent.[12] Although industry analysts project little agricultural pesticide sales growth in the industrial world, Latin America and developing Asia are growth markets,

expanding at between 3 and 6 percent a year as farmers in these nations rely increasingly on farm chemicals.[13]

The regional breakdown of agricultural sales depends not only on the crops being grown (and the prevailing diseases of those crops), but also on climate and the structure of food production. For instance, the highly mechanized farms of North America depend on widespread herbicide use, a situation reinforced by the recent introduction of crops genetically engineered to tolerate spraying; this region accounts for 40 percent of global herbicide sales.[14] On the other hand, insecticide and fungicide use are concentrated in the tropics and subtropics, where warm, wet conditions exacerbate insect and fungus outbreaks.[15]

Five crops—rice, corn, wheat, cotton, and soybeans—account for over half of all agricultural pesticide sales.[16] As a group, fruits and vegetables account for another quarter of use.[17]

At over $14 billion, herbicide sales capture nearly half of the global agricultural pesticides market.[18] Insecticide sales are nearly 30 percent of the market, while fungicides are nearly 20 percent.[19]

Because companies are reluctant to report quantities of pesticides sold, statistics on total use are not widely available; moreover, different compounds can vary in application rate by a factor of 100, making year-to-year comparisons difficult. Still, global use has been esti-

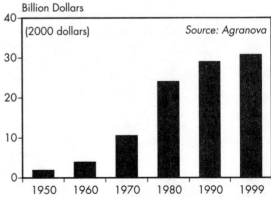

Figure 1: Global Pesticide Sales, 1950–99

mated at between 2.6 million and 3.5 million tons per year, up roughly 10-fold since 1950.[20]

Two companies—Syngenta (formed by the merger of the agrochemical divisions of Novartis and AstraZeneca) and Pharmacia (which recently absorbed Monsanto)—control 34 percent of the global agricultural chemical market.[21] The top 10 firms, all based in Europe, the United States, and Japan, control 84 percent of the global market.[22] Nearly all these companies are also heavily invested in biotechnology, reinforcing the use of farm chemicals in conjunction with genetically engineered seed.[23]

Worldwide, pesticide exports have grown from $1.3 billion (in 2000 dollars) in 1961 to over $11.4 billion in 1999.[24] Because national pesticide legislation varies around the world, countries will sometimes export compounds banned domestically. Of the 1.45 billion kilograms of pesticide exported by the United States—the largest exporter after Germany and France—between 1997 and 2000, nearly 30 million kilos were either forbidden or severely restricted in the United States.[25]

The World Health Organization estimates that every year 3 million people suffer from severe pesticide poisoning, matched by a greater number of unreported, mild cases that result in acute conditions such as skin irritation, nausea, and diarrhea.[26] And as many as 220,000 people die from pesticide poisoning.[27] For many of these compounds, the long-term health effects are poorly understood or unknown, particularly in the combinations of chemicals that humans usually encounter.

The U.N. Food and Agriculture Organization recently estimated that more than 500,000 tons of obsolete pesticides—old and unused pesticides that have been banned—have been stockpiled in the developing world and the former Soviet Union.[28] These stocks, which contain some of the most dangerous chemicals ever produced, threaten the health of millions of people, leak into the environment, and contaminate farmland and drinking water.[29] The United Kingdom spends roughly $200 million each year to remove pesticides from drinking water, equal to one quarter of what British farmers spend on pesticides themselves each year.[30]

Moreover, our pesticide use is becoming less effective as overuse has helped kill off all but the most resistant bugs.[31] For example, in the United States, the share of the harvest lost to pests has increased from 30 percent in the early 1940s to 37 percent in the 1990s—despite a 10-fold increase in pesticide use.[32] The annual cost of this resistance, in terms of additional crop losses and chemical expenses, is nearly $2 billion—equal to one fifth of what American farmers spend on pesticides each year.[33]

The health and environmental tolls of pesticide dependence, combined with the widespread emergence of resistant pests, have increased interest in organic farming and nonchemical approaches to pest control. Following farmer-field schools on insect ecology and nonchemical pest control, for example, and higher taxes on certain pesticides, 2 million farmers in Viet Nam have cut pesticide applications from 3.4 to just 1 per season, with no drop in yields.[34]

In Switzerland, where a radical restructuring of agricultural policy means that farmers must meet certain ecological standards in order to receive subsidies, pesticide use has fallen by one third in the last decade.[35] And in Ontario, Canada, a program to train vendors and users of pesticides in the basics of nonchemical pest control, pesticide safety, and pesticide regulations resulted in a 40-percent reduction in the amount of pesticides used on farms since 1987.[36]

# Resource Economics Features

Biotech Industry Growing

Appliance Efficiency Takes Off

Water Stress Driving Grain Trade

# Biotech Industry Growing

*Brian Halweil*

In every nation for which there are data, biotech revenues and investments are soaring. The combined annual revenues of biotech industries in the United States, the European Union, India, and Australia—which together account for the vast majority of the global industry—are now well over $35 billion.[1]

The broadest definition of the biotechnology sector, also known as the "life sciences" industry, encompasses any activities that use genetic engineering, DNA analysis, and other modern biological techniques for applications within medicine, agriculture, chemical manufacturing, and other industries.[2] (See Table 1.)

Links: pp. 26, 126, 138

Revenues of the U.S. biotech sector, which dominates the world market, increased from under $5 billion in 1989 to $25 billion in 2000.[3] The American biotech giant Amgen is almost as big as the entire European sector, the world's second largest market.[4] The total U.S. industry is likely much larger, since Ernst & Young, the source of these figures, defines the industry as companies strictly engaged in biotechnology, excluding chemical, pharmaceutical, or other companies that routinely use the technology.[5] In 2001, for example, U.S. sales of pharmaceuticals derived from biotech were valued at over $25 billion.[6]

The amount of money invested in U.S. biotechnology companies jumped from $45 billion in 1992 (the first year of significant investment) to $331 billion in 2000.[7] In contrast, $71 billion was invested in publicly held biotech companies in Europe in 2000.[8] Since the biotech index was launched by the American Stock Exchange in December 1994, it has outperformed the Nasdaq and the Dow.[9]

Biotech revenues in the European Union were roughly $8.2 billion in 2000, up more than fourfold since 1996.[10] The United Kingdom, Germany, France, and Switzerland have the leading biotech sectors there.[11]

Growth of the biotech industry in the developing world is more concentrated, with significant industries in Brazil, Cuba, and India. Cuba now generates about $100 million each year in revenues from vaccines, drugs, and other biotech products.[12] The biotech market in India is valued at $2.5 billion, up fivefold from 1997.[13]

A defining feature of this emerging industry is its seamless integration with pharmaceuticals, food, chemicals, cosmetics, agriculture, and other industries. For instance, bioengineered organisms and enzymes are being used in an array of industrial processes, from beer and cheese making to chemical manufacturing; most laundry detergents produced in the United States now contain genetically engineered enzymes.[14]

This phenomenon is particularly pronounced in the health sector, as pharmaceutical companies use biotech companies to expand research into new and potentially lucrative areas.[15] In 1998, drug companies signed research or licensing agreements with biotech firms worth $4 billion, a figure that rose to $7 billion by 2000.[16]

Medical products dominate global biotech sales. In the United States, more than 90 percent of annual sales is for medical applications, including pharmaceuticals and human diagnostics (disease test kits, for instance).[17] In contrast, just 5 percent of sales involve agricultural products, including transgenic seeds—the biotechnology that has attracted most public attention to date.[18] So far, biotech drugs have generated less public protest partly because drugs are taken voluntarily, they have a clear purpose, and they are not released into the environment.

Many aspects of the life sciences industry have inspired protest because they raise ethical issues about equitable access to medical innovations and the right to own living organisms. For instance, because the biotechnology industry depends so heavily on patents and other proprietary arrangements, various entities—from human genes to crop varieties—that were once considered the collective property of humanity are now the sole possession of private companies. The number of biotech patents—claims on snips of DNA, genetically engineered crops, or biotechnological processes—granted each year by the

U.S. Patent and Trademark Office grew sixfold between 1985 and 1999.[19]

An additional concern is raised by industry consolidation. The total value of mergers and acquisitions in the biotechnology sector jumped from $9.3 billion in 1988 to $172.4 billion in 1998.[20] The top five biotechnology firms, based in the United States and Europe, control more than 95 percent of the patents on gene transfer techniques.[21] Monsanto seed varieties account for 94 percent of the global area planted in transgenic crops.[22]

Perhaps the biotech applications that have generated the most controversy involve manipulating the very nature of human reproduction and human life. In February 2001, a private biotech firm and a publicly financed international consortium of scientists simultaneously described a rough map of the human genetic code, which could eventually yield major insights into human health and development.[23] Although the public effort's results are freely available, the private firm plans to patent and sell the discoveries in its map.

Many nations ban both human cloning and the engineering of humans with traits that could be passed on to future generations, although several teams of scientists have announced plans to clone a human being in the near future.[24] In November 2001, the United Nations called for a ban on human cloning.[25]

## Table 1: Various Applications of Biotechnology

Pharmaceuticals
From 1995 to 2000, nearly three times as many biotech drugs were approved in the United States as in previous 13 years. By mid-2000, biotechnology industry had over 14 percent of the products in medicinal trials there; share projected to increase to 25 percent by 2010. Four biotech drugs have at least $1 billion in annual sales: Epogen and Procrit (both for anemia), Intron A/Rebetron (a combination cancer treatment), and Humulin (a version of human insulin for diabetics).

Agriculture
Farmers planted genetically engineered (transgenic) crops on 52.6 million hectares in 2001, a 30-fold increase from 1996. Biotech being used to develop tree and other plant varieties, as well as livestock and fish, although no commercial products on the market yet. Scientists have genetically engineered pigs to make their organs less likely to be rejected by humans, and have engineered livestock to secrete human drugs in their milk or urine, which some think could significantly lower the cost of drug production.

Information Technology
"Bioinformatics" is the use of computers to make sense of biological data; today's computing power has already become indispensable for recording and analyzing the colossal information contained in genomes—whether by searching an individual's genes for a propensity for disease or a crop variety's genes for a tolerance to drought. Computer chips are being developed that use strands of DNA to do computations, an application that some believe will help shrink the size and increase the power of microchips.

Human Life
Scientists have suggested that embryonic stem cells—which form at an early stage of embryo development and have the ability to differentiate into any human cell—could help repair damaged human tissues and treat diseases. Many of the same techniques used to work with stem cells can be used to create a human clone or to genetically modify a human embryo—evoking images of mass-produced babies and creation of different genetic "classes." In late 2001, one company announced that it had begun cloning human embryos in an effort to generate stem cells.

Source: See endnote 2.

# Appliance Efficiency Takes Off

*Michael Scholand*

National energy standards and labeling programs are being initiated and expanded in a growing number of countries to save energy and other resources, such as water. The increasing global adoption of these programs is linked to the rapid uptake of domestic appliances in industrial and developing countries alike. Between 1994 and 1998, global sales of selected domestic appliances averaged 4.8-percent annual growth.[1] In China, in particular, appliance purchasing has grown dramatically. More than 65 percent of city-dwellers in China now own a refrigerator and more than 90 percent a clothes washer; both up from less than 5 percent only two decades ago.[2]

*Links:* pp. 38, 46, 124

Domestic appliances improve quality and convenience in our lives, but they also consume a great deal of resources. In part due to the increasing number of appliances, Chinese residential electricity consumption has grown on average 16 percent a year since 1985.[3] This has contributed to the addition of 16 gigawatts of new capacity every year to the Chinese grid—the equivalent of 32 medium-sized power plants.[4]

The energy and water consumed in running all these appliances need not be so great. If the market demands it, manufacturers will develop new and better products that perform the same services but have less environmental impact. In many countries, however, consumers have precious little information about the energy or water consumption of the appliances they buy. Effective labels are an essential first step in providing them with a tool to purchase more energy-efficient appliances. Minimum efficiency performance standards (MEPS) are an accompanying measure that eliminates the least efficient models from the market, benefiting consumers through lower operating costs.

Programs like these can now be found in 43 countries around the globe.[5] The number of nations implementing programs has increased sevenfold since 1980, and more than tripled in the past decade. (See Figure 1.) Nearly half the programs are found in Europe, and 14 are in Asia.[6] Europe's country count jumped in 1992, when the European Union (EU) launched a labeling and standards program that applied to all member states.[7]

Appliance labeling and minimum standards are two different yet complementary strategies for achieving the same goal—reducing energy consumption. Labels "pull" the market by providing consumers with information. The label can be either a recognizable quality certification or a comparative scale or rating scheme. These programs can be voluntary or mandatory, but they do not restrict any product from the market. Minimum efficiency performance standards, in contrast, "push" the market and are based on the concept of consumer protection. These standards are regulatory in nature, and often prohibit the sale of inefficient or poor-quality products.

For more than two decades, appliance labeling programs have sought to inform consumers about product characteristics that may not be readily apparent.[8] One example of this is the U.S. government's Energy Star program. Whether affixed to a refrigerator, a fax machine, or a light bulb, an Energy Star label helps shoppers identify efficient products with lower energy costs.[9] These products typically

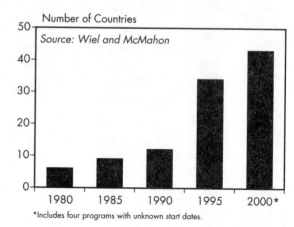

Figure 1: Countries with Appliance Efficiency Programs, 1980–2000

*Includes four programs with unknown start dates.

exceed federal efficiency standards by 13–20 percent, and sometimes by as much as 110 percent.[10]

Labeling programs like Energy Star provide one measure of merit, whether a product is compliant or not. Another major type of appliance label—comparative labels—provides a relative scale or ranking from which a consumer can compare energy consumption or some other performance characteristic. These scales take many forms, such as a one- to six-star rating (in Australia), a scale from A to G (in Europe), or a sliding bar showing relative energy consumption (in the United States and Canada).

In Thailand, appliance labeling programs have had a considerable impact on energy use. Comparative labels are credited with saving more than 100 megawatts of peak consumption to date. Estimates suggest an additional 200 megawatts of peak hour savings by 2005.[11] The Thai refrigerator program has transformed the single-door refrigerator market, increasing the market share of units rated five stars from 12 percent in 1996 to 96 percent in 1998.[12] The air conditioner program has been slower, but still doubled the market share of the most efficient air conditioners from 19 percent to 38 percent in the same period.[13]

After concluding that the U.S. EnergyGuide comparative label was difficult for some consumers to understand, the American Council for an Energy-Efficient Economy (ACEEE) launched a full-scale evaluation of the existing label and alternate designs.[14] Working with consumers, retailers, and manufacturers, ACEEE developed a comparative label using a one- to five-star rating that it believes is more consumer-friendly and should make an impact in the market. The U.S. government is expected to consider improvements to the EnergyGuide label in 2002.[15] The potential benefits of an improved label are clear: if 20 percent of American consumers were influenced to purchase one of the most efficient refrigerators available, for instance, the electricity savings would eliminate the need for more than four large power plants.[16]

The other essential tools for promoting more-efficient products, minimum efficiency performance standards, establish mandatory performance criteria either at a minimum level or a sales-weighted average. The first confirmed account of these were refrigerator standards adopted by the French government that took effect in 1966.[17] Over the years, MEPS have a demonstrated track record of success. In the United States, for example, they are credited with reducing the energy consumption of an average new refrigerator sold in 2001 by 75 percent compared with one manufactured 25 years earlier.[18]

MEPS usually target the lowest ranking products on the market. For example, the EU system of label categories (A to G) is referenced when setting a MEPS. In 1999, the EU eliminated the lowest-ranking refrigerators and freezers (E, F, and G) from the market.[19] From 1992 to 1999, the average energy use of cold appliances sold in the European Union fell by 27 percent, due to the energy label and manufacturers' improvements to meet the MEPS.[20] A review of this program found that for each euro spent on the efficient refrigerator program, 100,000 euros in electricity savings will be realized by EU citizens between 1995 and 2020.[21]

Recognizing the synergistic strengths of standards and labels, a global initiative has been launched to promote further global adoption of these programs—the Collaborative Labeling and Appliance Standards Program (CLASP). Initiated in 1999, CLASP's mission is to promote the appropriate use of energy-efficient standards and labels for residential and commercial appliances, equipment, and lighting in developing and transitional countries.[22] This program serves as both an information clearinghouse and a technical advice center, working globally to promote economic and environmental savings through energy efficiency standards and labeling.

# Water Stress Driving Grain Trade       *Sandra Postel*

As per capita water supplies drop below the level where it is possible or practical for countries to grow all their own food, numerous countries are satisfying their food demands by importing more grain. Collectively, water-stressed nations in Africa, Asia, and the Middle East now account for 26 percent of global grain imports—a figure that is likely to rise in the coming decades.[1]

Countries are classified as "water-stressed" when their per capita renewable water supply drops below 1,700 cubic meters per year. Typically countries in this situation do not have enough fresh water to meet the food, industrial, and domestic needs of their people.[2] Rather than importing water directly, they adapt to water stress by importing more of their food.

Links: pp. 26, 34, 60

It takes roughly 1,000 cubic meters of water—and considerably more than this in arid climates—to produce one ton of wheat.[3] Since grain is easier and more cost-effective to transport in large quantities than water is, water-strapped countries turn to world markets for a portion of their grain. They use their limited fresh water to support urban and industrial activities, which typically generate 50–100 times more economic value than the same quantity of water used in agriculture. The wealth-creating potential of this strategy has enabled some very water-scarce nations—Israel, for example—to achieve high standards of living.

The 36 nations in Africa, Asia, and the Middle East that are now categorized as water-stressed collectively import more than 68 million tons of grain a year.[4] With annual world grain imports averaging 260 million tons in recent years, these countries currently account for more than one quarter of global grain imports.[5] More than 20 water-stressed countries now rely on imports for at least one fifth of their grain consumption—and 15 use them for more than half. (See Table 1.)

In order for a strategy of relying on grain imports to work, at least two conditions must be satisfied. First, there must be enough surplus grain offered for export in world markets

to meet the demands of the importers. Second, that grain must be available at a price the importing nations can afford.

There is good reason to be concerned about whether these two conditions will be met in the future. By 2015, seven more countries—including Ethiopia, Iran, and Nigeria—will join the ranks of the water-stressed.[6] Adding the projected populations for these seven to the population increases expected in countries already water-stressed yields the unsettling conclusion that by 2015 the number of people living in water-stressed countries will grow by more than 800 million—to nearly 3 billion, or 40 percent of projected world population.[7] As water stress deepens and spreads, more countries will be driven into the camp of net grain importers and those already in this camp will be driven by population and consumption growth to import more grain.

Just over the last five years, a number of water-stressed countries have increased sharply their dependence on grain imports. Morocco's import dependency (net imports as share of consumption) climbed from an average of 26 percent between 1994 and 1996 to an average of 51 percent between 1998 and 2000.[8] At the same time, Algeria's import dependency rose from an average of 70 percent to 77 percent, while Saudi Arabia's climbed from 50 percent to 73 percent and Yemen's from 66 percent to 78 percent.[9] Syria was actually a small net exporter of grain in the mid-1990s, but became dependent on imports for an average of 23 percent of its grain consumption in the 1998–2000 period.[10]

Moreover, three large countries that currently produce most of their own food—China, India, and Pakistan—are likely to be driven by water stress and other factors to join the ranks of the grain importers in the near future. China already has severe water problems in its agriculturally important Hai and Yellow River basins. The projected 2025 water deficit for these two basins is roughly equal to the volume of water needed to grow 55 million tons of grain.[11] As much as one fourth of India's grain production—some 45 million tons—is jeopar-

## Table 1: Grain Import Dependence of Water-Stressed Countries in Africa, Asia, and the Middle East, Circa 2000

| Country | Net Grain Imports as Share of Consumption[1] (percent) |
|---|---|
| Kuwait | 100 |
| Oman | 100 |
| United Arab Emirates | 100 |
| Lebanon | 97 |
| Israel | 96 |
| Jordan | 91 |
| Libya | 89 |
| Yemen | 78 |
| Algeria | 77 |
| Saudi Arabia | 73 |
| South Korea | 71 |
| Iraq | 63 |
| Mauritania | 60 |
| Tunisia | 53 |
| Morocco | 51 |
| Egypt | 39 |
| Azerbaijan | 39 |
| Somalia | 37 |
| Kenya | 33 |
| Syria | 23 |
| Ghana | 21 |

[1]Ratio of annual net grain imports to grain consumption averaged over 1998–2000. Includes 21 of the 36 water-stressed countries importing at least one fifth of their grain.
Source: Global Water Policy Project, based on population data from U.S. Census Bureau and grain data from U.S. Department of Agriculture.

other factors that also cause nations to rely more heavily on imported grain.[15] As a result, the pursuit of food security by trading other goods and services for "virtual water"—perhaps a wise strategy for each individual water-stressed nation—may not be so wise when applied to all nations in this situation.

In the absence of an international food aid bank or other global mechanism for filling food supply gaps, this may indeed be a risky strategy for poorer water-stressed countries that do not have the foreign-exchange earnings to handle large fluctuations in world grain prices. The vast majority of the increase in water-stressed populations will occur in sub-Saharan Africa and South Asia, sites of the deepest pockets of hunger and poverty today.

Sound strategies for coping with water stress need to include serious efforts to conserve water and use it more efficiently. Much water continues to be wasted or used unproductively even in water-scarce regions. In addition, ensuring that water scarcity does not translate into more hunger requires efforts well beyond a grain-import strategy. Adequate levels of grain per person—whether achieved through domestic production or imports—does not alleviate hunger and malnutrition unless the hungry can afford to buy food or otherwise get access to it. Improving the food-producing capabilities of poor farmers directly—through the spread of low-cost irrigation, for example—is a surer way of reducing hunger.[16]

dized by groundwater overpumping alone.[12] Thus India and China could be headed toward combined grain imports of 100 million tons—more than the entire current U.S. supply of grain to world markets.[13] Pakistan is plagued by shortages of Indus River water, groundwater overpumping, and serious salinization of its irrigated lands, and seems unlikely to remain food self-sufficient for long.[14]

All in all, water stress will become a much bigger driver of the international grain trade in the coming years. This will occur against a backdrop of land scarcity, urbanization, and

# Health Features

© DIGITAL VISION

Food-borne Illness Widespread

Soda Consumption Grows

Prevalence of Asthma Rising Rapidly

Mental Health Often Overlooked

# Food-borne Illness Widespread

*Danielle Nierenberg*

Food-borne illness is one of the most widespread health problems worldwide, and it could be an astounding 300–350 times more frequent than reported.[1] (See Table 1.) Although food-borne diseases strike up to 30 percent of the population in industrial nations each year, the infections are usually minor and easily treatable.[2] Developing nations bear the greatest burden of this problem because of the presence of a wide range of parasites, toxins, and biological hazards and the lack of surveillance, prevention, and treatment measures that ensnarl the poor in a chronic cycle of malnutrition and infection.[3]

Links: pp. 24, 28, 60

The World Health Organization estimates that more than 1.5 billion episodes of diarrhea occur each year in children under the age of five from ingesting tainted food and water, leading to more than 3 million deaths.[4] For all ages, experts believe that 70 percent of diarrheal disease may be caused by food.[5] Cholera can be spread by contamnated food or water; in 1997, 65 nations reported outbreaks leading to 6,000 deaths.[6]

In the United States, food contaminated with bacteria, parasites, fungi, and viruses causes some 76 million illnesses, 325,000 hospitalizations, and up to 5,000 deaths annually, but because of underreporting the true figure is likely much higher.[7] The United Kingdom has experienced a rapid increase—perhaps as high as fourfold—of reported food-borne illness in the last two decades, and similar trends are evident in Australia, Canada, Germany, Japan, New Zealand, the Netherlands, Sweden, and other nations that track food-borne disease.[8]

Globalization, human migration, changes in eating habits, agricultural technologies, and environmental change can all influence the rise and spread of food-borne disease. The combination of consumer demand for fresh fruits and vegetables year round and new trade routes means food is travelling longer distances to markets.[9] Increased consumption of food bought from street vendors—and poor hygiene practices—help spread food-borne illnesses in the urban developing world.[10] Hepatitis A can be spread by food handlers—an epidemic in China in the 1980s struck some 300,000 people.[11]

It can be difficult to determine the primary sources of food-borne hazards. In the United

## Table 1: Selected Pathogens

| Pathogen | Description |
| --- | --- |
| Camplyobacter | Half of infections are associated with eating contaminated poultry or handling chickens; the most common food-borne infection in the United States. |
| Listeria | Present in soft cheese and meat pastes; for healthy adults it may cause no symptoms at all, but among pregnant women, infants, the elderly, and the ill, the death rate is about 30 percent. |
| Marine Toxins | Poisonous marine life cause almost 45 percent of known food-borne disease in the Caribbean and Latin America. Just one—*Ciguatera*—affects 50,000 people annually. |
| Parasites | Amoebas—parasites spread by contaminated food and water—cause 100,000 deaths a year, second only to malaria in mortality due to parasites; 10 percent of the world is at risk of contracting trematodes, parasites found in raw freshwater fish, shellfish, and aquatic plants. |
| Pathogenic *E. coli* | Responsible for up to 25 percent of all cases of diarrhea among children and infants in the developing world; caused by consumption of food that has come into contact with fecal matter. |
| Salmonella | Spread primarily through raw or undercooked eggs, poultry, and milk; accounts for the greatest proportion of food-borne disease in industrial countries. |

*Source:* See endnote 1.

States, most of these illnesses are a result of eating contaminated fish, shellfish, fruits, or vegetables, followed by meat and poultry products.[12] In the United Kingdom, the culprit is meat and eggs.[13] And in Cuba, the Dominican Republic, El Salvador, and Haiti, fish, water, and red meat are the top three major sources.[14] Animal manure and human fecal matter are both common and growing contaminants of meat products and of fresh fruits and vegetables that are stored, processed, or shipped in unsanitary conditions.

Traditionally, consumers have been blamed for food safety problems—accused of having poor personal and household hygiene, undercooking food, or not storing crops and food properly. Aflatoxin (a type of mould), for instance, grows on crops kept in humid conditions. It can lead to both fatal outbreaks of aflatoxicosis and high rates of liver cancer in some regions of Africa, Southeast Asia, and China.[15] In one Chinese village, adults have a 1 in 10 chance of contracting liver cancer from eating contaminated grains.[16] Botulism results from improper canning, and mortality rates are between 35 and 65 percent.[17]

But many food safety problems start long before they reach the consumer. According to the U.N. Food and Agriculture Organization, the trend toward increased commercialization and intensification of livestock leads to a variety of food safety problems.[18] Crowded, unsanitary conditions and poor waste treatment in factory farms exacerbate the rapid movement of animal diseases and food-borne infections. Samples of water downstream of one facility in Michigan contained 1,900 times the state's maximum standard for *E. coli* in surface waters, and over 1, 000 people were sickened by *E. coli* in Walkerton, Ontario, after the town's drinking water was polluted by nearby cattle operations.[19]

Livestock feeds rich in starch but lacking hay—the standard diet in factory farms—have also been linked to the spread of food-borne pathogens, such as *E. coli*.[20] Factory-farmed poultry can disperse salmonella widely in the environment, polluting surface waters, the soil, and rivers.[21] Heat-induced stress during the summer increases susceptibility to illness in factory farms, where animals are bred for their reproductive potential.[22] Unfortunately, antimicrobial drugs mixed into the feed and water of cattle, chickens, and hogs to prevent disease have increased antibiotic resistance in bugs found in livestock and humans alike, making it increasingly difficult to battle emerging pathogens.[23]

New technologies in agriculture, including pesticides, agricultural chemicals, and, more recently, genetic engineering, have the potential to create food safety disasters. Bovine spongiform encephalopathy (mad cow disease), a virus caused by feeding cattle the renderings of other ruminants, can be spread to humans who eat infected meat. To date, more than 100 people in the United Kingdom, France, and Ireland have died from new variant Creutzfeldt-Jakob disease, the human form of mad cow.[24] Some experts predict that the number of victims could top 100,000 by mid-century.[25]

Up to 200 Japanese sushi aficionados are poisoned each year by a toxin found in pufferfish; the mortality rate can be as high as 50 percent.[26] And the tradition of eating bush meat in Gabon has helped spread the Ebola virus—a disease that can be spread by eating infected meat from primates and has a mortality rate of 90 percent.[27]

Food-borne diseases have tremendous economic and societal costs. They are the most frequent reason children are hospitalized in developing countries, and they lead to increased absenteeism from school and work, loss of income, and poor productivity.[28] In the United States, the costs of just seven food-borne pathogens range between $5.6 billion and $9.4 billion annually.[29] An epidemic of cholera in 1991 cost Peru more than $770 million in lost exports, decreased tourism, and food service closures.[30]

# Soda Consumption Grows
*Erik Assadourian*

In 2000, the global consumption of carbonated soft drinks (soda) reached 179 billion liters—29.4 liters per person.[1] (See Figure 1.) Soda maintained its ranking as the third most popular commercial beverage and edged closer to milk, which fell to 196 billion liters (32.2 liters per person).[2] While milk consumption fell 3.0 percent between 1999 and 2000, soda consumption grew 2.9 percent.[3]

The United States, with less than 5 percent of the world's population, is the largest soda consumer and accounted for one Link: p. 32 third of total soda consumption in 1999.[4] (See Table 1.) The 58 billion liters sold there generated $48 billion dollars in revenue for the soda industry.[5] Soda is already the number one drink for Americans, who took in an average of 211 liters of it in 1999—compared with 109 liters of tap water.[6]

This rapid growth in soda consumption is also occurring in the developing world. China, with about a fifth of the world's population, is the fourth largest consumer of soda.[7] Between 1994 and 1999, per capita consumption in China grew 60 percent, to 7 liters per year.[8] Annual per capita consumption in Brazil, the third largest soda market, also shot up 60 percent between 1994 and 1999, reaching 61 liters per person.[9]

Unlike juices or milk, which contain vitamins and important minerals like calcium, soda consists of carbonated water, sweeteners (either caloric or high-intensity), flavoring, and in many cases caffeine. Consumption of these calorie-dense but nutritionally devoid drinks often displaces healthier foods, which can lead to dietary deficiencies.[10]

In the United States, as soda consumption doubled between 1970 and 1999, milk consumption fell 25 percent.[11] During this period, total calcium intake by children fell significantly.[12] A recent study found that children who drank soda took in a significantly smaller amount of vitamin A and calcium each day than those who drank milk.[13] As calcium is central to building strong bones, and as most bone mass in women is built by age 18, an increase in osteoporosis rates is a real threat.[14] A recent preliminary study found that drinking soda is significantly associated with increased prevalence of bone fractures in active adolescent women.[15]

As soda is a large source of added sugars and calories, it can also contribute to obesity. A recent study showed a direct correlation between consumption of sugar-sweetened drinks and childhood obesity.[16] The results suggested that children increase their odds of becoming obese by 60 percent with each additional sugar-sweetened drink they consume.[17] In America, overweight and obesity among children have tripled to 14 percent since 1970, and have increased to 61 percent among adults.[18] On average, Americans consumed about 185 calories from soda each day in 1999, which is more than the suggested daily maximum of added sugars.[19]

Soda consumption can also contribute to tooth decay. Although all sugars can cause tooth decay, soda is a primary concern because it is often consumed between meals or sipped over a long period, which prolongs the time that sugars remain in the mouth.[20]

Of the top 10 global brands of soda, more than 80 percent of the volume sold in 1999 contained caffeine.[21] This mood-altering drug

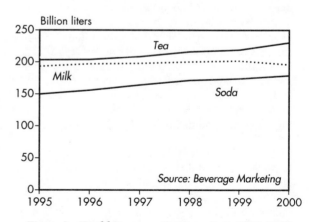

Figure 1: World Beverage Consumption, 1995–2000

is physiologically and psychologically addictive and can produce physical dependence with a daily intake of just 100 milligrams.[22] Coca-Cola, the world's most popular brand, contains 34 milligrams of caffeine per 355-milliliter can.[23] Because the effects of caffeine are weight-proportionate, a child will be more strongly affected by a small amount of caffeine.[24] While caffeine is supposedly added to enhance soda's flavor, a recent study found that only a small percentage of consumers were able to tell the difference between caffeinated and caffeine-free colas.[25]

The soda industry aggressively markets its products. In 2000, the two largest soft drink corporations, the Coca-Cola Company and PepsiCo., spent $4.6 billion worldwide on advertising.[26] A significant portion of this directly targets children, often connecting soda with children's heroes. For example, Coca-Cola signed an exclusive $150-million global contract with Warner Brothers, the producer of *Harry Potter and the Philosopher's Stone,* to be the sole marketing partner for the movie.[27]

The soda industry also markets to children in schools, often signing exclusive marketing contracts with school boards, which in many cases tie monetary bonuses to a minimum amount of soda sold. In response, some schools have ended contracts after community objections.[28] In early 2001, Coca-Cola announced that it would start selling more nutritious beverages along with soda in U.S. schools.[29] Yet this change is probably as motivated by economic considerations as by grassroots pressure—recognizing that the U.S. soda market is saturated, Coca-Cola has started to diversify its product base to include other soft drinks such as water, juices, and sports drinks.[30]

With obesity becoming a global epidemic, health organizations and governments are trying to encourage healthier diets and lifestyles.[31]

## Table 1: Market Share and Per Capita Consumption of Carbonated Soft Drinks, Top Five Countries, 1999

| Country | Share of Global Market (percent) | Per Capita Consumption (liters) | Growth Per Capita, 1994–99 (percent) |
|---|---|---|---|
| United States | 33 | 211 | 10 |
| Mexico | 8 | 146 | –3 |
| Brazil | 6 | 61 | 60 |
| China | 5 | 7 | 60 |
| Germany | 4 | 92 | 18 |
| Top Five | 57 | 53 | 15 |

Source: Beverage Marketing, *The Global Beverage Marketplace, 2001 Edition* (New York: 2001).

In a recent campaign, the Washington-based Center for Science in the Public Interest mobilized the health and education communities to "Save Harry Potter" from Coca-Cola and prevent children from being the target of an aggressive advertising campaign.[32]

Several countries have restricted the marketing of products to children. In Poland, for example, there is a ban on all television and radio marketing to children, which has significantly reduced product sales, including of soda.[33] Sweden also bans advertising to children on TV. But because of the strong presence of satellite TV, to which the ban does not apply, this has had less impact on consumption.[34]

In the United States, several states tax soda and other "junk foods." California, for example, has a 7.25-percent sales tax on soft drinks, which results in an annual revenue of $218 million.[35] Junk food taxes help reduce consumption of these unhealthy, often packaging-intensive foods and beverages. Further, while these taxes currently go to general funds, using them to counteract the huge advertising budgets of the soda and other junk food industries would help counter their pervasive messages and educate consumers about the importance of a healthy diet.

# Prevalence of Asthma Rising Rapidly       *Uta Saoshiro*

Worldwide, an estimated 100–150 million people suffer from asthma today.[1] On average, prevalence rates have been rising by 50 percent every decade.[2] As one of the most common chronic diseases, asthma is increasingly being recognized as an international health problem.[3] Though mortality due to asthma (at 218,000 in 2000) is significantly less than major killers such as tuberculosis (1.7 million), there are substantial social and economic costs related to asthma.[4] Estimates of total family income spent on medical treatment range from 5.5–14.5 percent in the United States to 9 percent in India.[5] It is also a major cause of school absences for children and lost work-days for adults.[6] In 2000, the direct costs (hospitalization, physician and nursing care, and medication) and indirect costs (lost workdays for adults and lifetime earnings lost due to mortality) together in the United States were estimated at $12.7 billion.[7]

*Link:* p. 148

Asthma is a chronic inflammatory lung disease, which makes a person much more sensitive to a variety of irritants and allergens.[8] When exposed to these, the airways constrict, causing breathlessness, chest tightness, and wheezing.[9] The symptoms, frequency, and severity of these attacks can be either mild or life-threatening, depending on the individual.[10] Asthma often begins in childhood, when immunity systems are still developing and vulnerable.[11] Although about one in four children eventually outgrows the disease, episodes early in life mean that a person is more likely to have recurring asthma as an adult.[12]

In mild cases, a person could have attacks once a week, sometimes limiting physical activity and sleep. In its severe form, asthma can be completely debilitating, characterized by continuous symptoms, frequent attacks, and significant limitation of routine activities—even walking up stairs.[13] Factors that exacerbate asthma include viral infections such as influenza, physical exercise, breathing cold air, laughing or crying hard, certain drugs (for aspirin-sensitive asthmatics), and summer hay fever.[14]

The first global study of asthma in children is being conducted in 56 countries.[15] So far, the International Study of Asthma and Allergies in Childhood (ISAAC) found the highest percent of children with the disease in Australia, New Zealand, the United Kingdom, and the United States.[16] (See Table 1.) These countries also had a high number of people with asthma among adults aged 20 to 44.[17]

The prevalence of asthma for the United States has more than doubled in just 16 years—from 6.8 million people in 1980 to 17.3 million in 1998, with deaths also doubling, to 5,637 in 1995.[18] The number of asthmatics in Europe has also doubled in the past 10 years.[19]

Although asthma is viewed as a problem primarily in industrial countries, an estimated 15–20 million people in India suffer from it as well.[20] Children are particularly hard hit, with 10–15 percent of asthma in India occurring in children between the ages of 5 and 11.[21] In Kenya, an estimated 20 percent of children have asthma, and the number varies from 20 to 30 percent in Brazil, Costa Rica, Panama, Peru, and Uruguay.[22]

The global rise in asthma is "one of the biggest mysteries in modern medicine," says the World Health Organization (WHO).[23] The causes of the disease are complex. So far the scientific community agrees that asthma results from a combination of genetic, environmental, and other factors.[24] There is strong evidence that some people are predisposed to asthma because one or both parents have the disease.[25] But researchers do not yet know which gene makes it more likely a child will develop asthma.[26]

Repeated exposure to indoor allergens, such as dust mites, furry pets, cockroaches, and molds, is one of the biggest risk factors, especially in infants. Outdoor allergens such as pollens and fungi can also be a problem. These and other allergens sensitize a person's airways to irritation and can cause asthma.[27]

Chemical irritants and allergens in the workplace are strong risk factors for adults.[28] The dangerous agents include everything from flour for bakers to disinfectants for hospital workers.[29] Exposure to passive smoking in the workplace is also a strong contributing factor,

## Table 1: Asthma in Children, by Region

| Region | Estimated Prevalence (percent) |
|---|---|
| Oceania | 25.9 |
| North America | 16.5 |
| Latin America | 13.4 |
| Western Europe | 13.0 |
| Eastern Mediterranean | 10.7 |
| Africa | 10.4 |
| Pacific Asia | 9.4 |
| Southeast Asia | 4.5 |
| Eastern Europe | 4.4 |

Source: See endnote 14.

and has consistently been found to exacerbate asthma.[30] This is also the case for children exposed to parental smoking, especially by the mother—the development of smaller lungs is typical of infants of mothers who smoke, thereby increasing the risk of the child developing asthma.[31]

Air pollution, on the other hand, has yet to be established as a causal factor, although it is an important contributing factor, according to the ISAAC study.[32] In areas with high levels of sulfur dioxide and particulate matter pollution (emitted mainly by coal burning for power and heating), such as China and Eastern Europe, the ISAAC study found a low rate of asthma among children (although the general prevalence of and mortality from other respiratory diseases, such as chronic obstructive pulmonary disease, are much higher).[33] Yet high rates in children were found in countries like New Zealand that have relatively low levels of air pollution.[34]

WHO estimates that 30–40 percent of asthma may be linked to air pollution in some populations.[35] But the precise role of different types of air pollutants remains ambiguous. Different studies show that high levels of ozone and nitrogen oxide pollution can exacerbate existing asthma, while studies relating sulfur dioxide pollution to asthma are not as clear.[36] Studies have also not controlled for other risk factors such as indoor allergen exposure, making it uncertain to what extent ambient air pollution affects asthma prevalence.[37] Most recently, results from a 10-year study of children in southern California suggested that ozone pollution not only exacerbates asthma, but can also cause the disease in children.[38]

Other factors, such as socioeconomic disadvantage, are associated with the higher prevalence of asthma. From inner-city America to the urban slums of Nairobi, Lagos, and Kinshasa in Africa, the prevalence of asthma is much higher—for example, adult prevalence in these three African cities ranges from 7 to 10 percent, but is 15–20 percent in their slums.[39] Factors such as inadequate waste disposal, poor housing conditions, and lack of access to proper medical care contribute to this state of affairs.

Fortunately, an estimated 95 percent of asthma can be controlled by continuous medical care.[40] In 1999 WHO added an anti-inflammatory drug, beclomethasone, to its essential drugs list (which tells countries the safe drugs for treating diseases affecting the majority of the world).[41] Even the most basic drugs are either unavailable or unaffordable in many developing countries, however.[42]

Prevention measures such as not smoking, forgoing carpets and pets, or decreasing involuntary exposure to secondhand smoke can all improve overall respiratory health.[43] A recent initiative in North Lanarkshire in the United Kingdom found that asthma patients' visits to the doctor dropped by two thirds when conditions favorable to house mites were eliminated by steaming carpets, renewing and cleaning bedding often, and installing better ventilation.[44] Although the exact impacts of various air pollutants remain unclear, a precautionary approach would suggest the enforcement of clean air rules, as well as keeping children indoors on high pollution days.[45]

# Mental Health Often Overlooked

*Danielle Nierenberg*

In 2001, the World Health Organization (WHO) estimated that 450 million people worldwide had a mental or neurological disorder.[1] Twenty-five percent of the population can expect to experience one or more disorders within their lifetimes.[2] Mental illness is universal, affecting people in all nations and from every background, but poor people in developing countries lack access to many of the most basic resources for effective treatment.

WHO's definition of mental health disorders is broad, encompassing a wide range of problems of both the mind and brain.[3] (See Table 1.) It includes autism, Alzheimer's disease, schizophrenia, depression, sleep disorders, addiction and substance abuse, bipolar affective disorder, panic and anxiety disorders, mental retardation, and epilepsy.[4] (Although epilepsy occurs because of an electrical mix-up in the brain and retardation and autism are developmental problems, people with these conditions are often discriminated against and prevented from fully participating in normal social activities.)

Links: pp. 148, 156

Overall, mental disorders account for almost a third of global disability (the number of healthy life years lost to a disability) from all diseases.[5] Depression is by far the most debilitating—more than 120 million people are affected worldwide.[6] Currently, depression represents 12 percent of the global disability burden, and by 2020 its share is expected to rise 15 percent, second only to heart disease.[7]

Although the incidence of depression is highest during middle age, experts recognize that the elderly and children are not immune to mental health problems. The prevalence of some disorders—dementia and Alzheimer's—rises with age.[8] In the United States, 1 in 10 young people suffers from impairment of psychological development or from behavioral, emotional, and depressive disorders.[9] Roughly 18 percent of children and adolescents in Ethiopia have a mental disorder, while in India the figure is 13 percent.[10] More than 20 percent of young people in Germany, Spain, and Switzerland are afflicted with depression, anxiety, or other mental problems.[11]

Rural isolation and poverty can make things

## Table 1: Selected Mental Health Problems

| Disorder | Description |
| --- | --- |
| Depression | Twenty percent of cases never go into remission; recurrence rate after first episode is as high as 60 percent. |
| Schizophrenia | Found equally in women and men; affects 24 million people worldwide. |
| Substance abuse | Dependence on tobacco, alcohol, and illicit drugs affects millions of people and is a rising problem in developing nations. |
| Epilepsy | Caused by excessive electrical activity in the brain—not dementia—it affects about 50 million people worldwide. |
| Obsessive compulsive disorder | Characterized by uncontrollable anxious thoughts or rituals; more common than schizophrenia, bipolar disorder, or panic disorder and affects about 2 percent of the U.S. population. |
| Eating disorders | Between 5 and 20 percent of people with anorexia nervosa, a disease characterized an intense fear of weight gain, die as a result of complications. Other disorders, including bulimia nervosa and binge-eating, are becoming more common among young women and girls in non-western nations, such as Japan, Brazil, and South Africa. |

*Sources:* See endnote 3.

worse. In remote regions, mental and general health care facilities or counselors are nonexistent or too expensive. Rural women—who also suffer from economic hardship—are more than twice as likely to suffer from depression than the general population.[12] Often the mentally ill, who carry the extra burden of being poor, wind up incarcerated. In the United States, there are five times as many prisoners with mental illness as there are patients in state mental hospitals.[13]

Changing societal norms can also bring out psychological problems as people are separated from their traditional social safety nets of family and community. For instance, eating disorders—an increasingly common problem among girls (and more and more boys) in affluent nations—have spread to developing countries as cultural definitions of female beauty change.[14] Dependence on a cash economy, overcrowding, pollution, and increased violence in cities can also exacerbate mental disorders.[15]

Mental illness strikes men and women differently. Almost 10 percent of women have a depressive episode every year, compared with fewer than 6 percent of men.[16] Men, however, are more likely to have substance abuse problems and antisocial personality disorders.[17] Severe mental disorders, such as schizophrenia, show no clear gender preference.[18]

Mental illness often exacerbates and in some cases leads to other health problems. Patients with untreated mental disorders who also suffer from other chronic conditions, such as cancer, HIV/AIDS, heart disease, or diabetes, are less likely to experience an improvement in overall health.[19] And addiction to drugs, tobacco, or alcohol—which WHO also classifies as mental health disorders—can increase the severity and duration of mental illness. Studies show that the mentally ill are about twice as likely as others to smoke.[20] Alcohol abuse is on the increase in many of the world's developing regions, especially among indigenous groups, who previously had little exposure to intoxicants.[21]

Suicide is the most tragic outcome of mental illness. Nearly 1 million people end their lives each year and an estimated 10–20 million people try to kill themselves.[22] Suicide—usually preceded by severe depression or schizophrenia—is a leading cause of death in young adults (15–34 years of age) in China and most of Europe.[23] In the United States, farmers in the upper Midwest—a region plagued by economic hardship and the loss of small farms—are 1.5–2 times likelier than other groups of men to commit suicide.[24] There is a strong correlation between violence against women and contemplation of suicide.[25] WHO found that Japanese victims of domestic violence were more than 30 times as likely to commit suicide as women who were not abused.[26] Battered women in the United States are five times more likely to commit suicide.[27]

Available treatment methods for mental illness vary regionally and among socio-economic classes. Use of psychotropic drugs—mostly in industrial nations—is rapidly increasing. Antidepressants are the third most often prescribed drug, with sales of over $13 billion worldwide.[28] The number of Americans taking medicines to treat their depression has risen by more than two thirds over the last decade.[29] Unfortunately, many are not supplementing their drug therapy with counseling or other interactions with mental health professionals.

In developing nations, however, therapeutic drugs for mental illness are usually unavailable to the general population. As a result, many people end up hospitalized—often in crowded, unsanitary asylums where they are neglected and abused—for conditions that could be treated with drugs, therapy, or both. Human Rights Commissions in India and Central America found that at least one third of the "inmates" in these hospitals were people with epilepsy or retardation, who need not be hospitalized.[30]

Few nations have adequate mental health programs, and many lack even the most basic or rudimentary services. WHO recommends that all nations provide treatment for mental illness as part of primary health care, launch public awareness campaigns to break stereotypes about mental illness, support community care of affected individuals, develop the human resources necessary to treat mental health care, and support research on mental illness.[31]

# Social Features

Poverty Persists

Car-Sharing Emerging

Sprawling Cities Have Global Effects

Teacher Shortages Hit Hard

Women Subject to Violence

Voter Participation Declines

# Poverty Persists

*Molly O. Sheehan*

Even though average incomes more than doubled in developing countries between 1965 and 1998, 1.2 billion people—more than one in five in the world—lived on less than $1 a day in 1998, a level used by the World Bank to denote "extreme poverty" or lack of income to meet basic food needs.[1] (See Table 1.) Although the share of people in this category fell between 1987 and 1998, the total number remained almost constant as population surged.

Roughly 70 percent of people surviving on less than $1 per day live in sub-Saharan Africa and South Asia.[2] Sub-Saharan Africa has the largest share of extremely poor people, although there are tremendous differences within and between nations.[3] The AIDS epidemic, which disproportionately kills people between the ages of 15 and 49, has worsened poverty in this region, as the disease takes the main wage earners in many families.

Links: pp. 88, 142, 144, 154

Rural areas house the bulk of the extremely poor, but the rural-urban income gap has been shrinking.[4] In Nigeria, for example, Africa's most populous nation, a failing economy and massive migration toward cities in the 1990s meant urban poverty outpaced rural poverty.[5]

Eastern Europe and Central Asia had the largest percentage increase of people living in poverty in the 1990s, following the breakup of the Soviet Union and the collapse of centrally planned economies. In Russia, the share of people living beneath the nationally defined poverty line surged from 11 percent during the Soviet era to 43 percent in 1996.[6]

Many thresholds of poverty exceed the $1-a-day measure. In industrial countries, some 130 million people live in poverty as defined by earning half of national median income.[7] In a recent book, a U.S. reporter chronicles three futile attempts from late 1998 to 2000 to maintain her health and dignity while earning $6–7 an hour, which is what some 34 million people living below the U.S. poverty line try to do.[8]

Poverty is about more than income: education and health reveal important distinctions between rich and poor.[9] Despite progress in the last few decades, 854 million adults are still illiterate, 2.4 billion people lack basic sanitation, and every day 30,000 children under the age of 5 die of preventable causes.[10] The Human Development Index produced by the U.N. Development Programme (UNDP) includes adult literacy, educational enrollment, and life expectancy. Most countries improved their ranking on this index between 1975 and 2000, but 20 nations in Africa, Eastern Europe, and the former Soviet Union actually fell backward on this scale.[11]

Poor people are also disproportionately vulnerable to environmental risks, crime, and government corruption. One fifth of the total burden of disease in the developing world is caused by environmental risks—from lack of safe water to exposure to industrial chemicals.[12] When World Bank researchers consulted more than 60,000 poor men and women in 60 countries, many voiced frustration over their powerlessness to protect their families from threats like floods, disease-carrying mosquitoes, thieves, and crooked officials.[13]

Not only has poverty persisted in the 1990s, but the chasm between rich and poor has widened.[14] While three fifths of the world's people earn just 6 percent of the world's income, one sixth receives 78 percent.[15] The gap between rich and poor shrunk in many countries between the 1950s and mid-1970s, but the reverse has happened since then.[16] Of 73 countries for which good data exist, inequality rose between the 1970s and 1990s in 48 nations, including the United States, most of Latin America, Russia, most of the former Soviet bloc, China, and parts of Africa.[17]

Extreme poverty in an era of unprecedented wealth is not merely shameful—it is dangerous. Many recent studies have shown a correlation between more equal income distribution and better public health, with life expectancy higher in countries like Sweden, Norway, and Japan, where the poorest households receive a higher share of income than in other wealthy nations.[18] Comparing U.S. metropolitan areas, researchers found a higher level of premature deaths in the places with the highest inequality.[19] Poverty and inequality can also

## Table 1: People Living on Less Than $1 a Day, Selected Regions, 1987 and 1998

| Region | 1987 | | 1998 | |
|---|---|---|---|---|
| | Total | Share of Population | Total | Share of Population |
| | (million) | (percent) | (million) | (percent) |
| Sub-Saharan Africa | 217.2 | 46.6 | 290.0 | 46.3 |
| South Asia | 474.4 | 44.9 | 522.0 | 40.0 |
| Latin America & Caribbean | 63.7 | 15.3 | 78.2 | 15.6 |
| East Asia & Pacific | 417.5 | 26.6 | 278.3 | 15.3 |
| Eastern Europe & Central Asia | 1.1 | 0.2 | 24.0 | 5.1 |
| Middle East & North Africa | 9.3 | 4.3 | 5.5 | 1.9 |
| Total | 1,183.2 | 28.3 | 1,198.9 | 24.0 |

Source: World Bank, World Development Report 2000/2001 (New York: Oxford University Press, 2000), p. 23.

lead to political instability and social tensions that impede economic growth and spark fanaticism and violence.[20]

Many nations are failing to meet international goals to reduce poverty.[21] In 2000, world leaders meeting at the United Nations for a Millennium Summit committed to several laudable goals for 2015: halving the number of people living in extreme poverty, suffering from hunger, and living without safe water; reducing infant mortality by two thirds; and enrolling all children in primary school. As of late 2001, 74 countries—with more than one third of the world's population—were not on track to halve poverty, and 83 countries, home to 70 percent of the world, were not on schedule to halve their share of people without access to safe water.[22]

While many attempts to reduce poverty have focused on stimulating national economies, economic growth alone does not guarantee cuts in poverty or inequality. A study of 47 developing countries between the 1980s and 1990s identified 117 growth periods; in 30 percent of these, the rapid growth did not affect poverty levels.[23] China's coastal cities have gained from rapid economic growth in the past 20 years, but rural inland communities have not—and inequality has soared as a result.[24] In the United States, the economic boom of the 1990s did not dent the nation's historically large income gap, even in prosperous metropolitan areas.[25]

Improving the health and education of women and girls can help reduce poverty. Although global breakdowns of poverty by sex are difficult to make, UNDP estimated in 1995 that women may account for as much as 70 percent of the world's poor.[26] Some 64 percent of illiterate adults are women.[27] When women lack education and health care, population growth can overwhelm capacities of countries to invest in needed social infrastructure.[28] Greater economic opportunities arose in East Asian nations such as South Korea, Thailand, and Taiwan after population growth slowed, fostered by policies to promote family planning and the education of girls.[29]

Accountable governments can also help alleviate poverty.[30] Burdensome laws and corrupt officials thwart many from entering the "legal" economy. Even in the poorest slums, people have built their own homes and businesses, but their lack of legal title to property prevents them from getting bank loans to further their progress. More than half the population of large cities such as Cairo, Nairobi, and Bombay, for example, lacks legal residences.[31] Economist Hernando de Soto estimates the value of real estate not legally owned in the developing world and former Soviet bloc nations at $9.3 trillion, all of which could be brought into the legal system and leveraged to reduce poverty.[32]

# Car-Sharing Emerging

*Gary Gardner*

Car-sharing, a subscription-based transportation service that makes cars available to its members, is taking off rapidly in industrial nations as environmentally conscious entrepreneurs tap the market for an alternative to owning or renting an automobile. Some 126,000 people share nearly 5,500 vehicles in car-sharing organizations (CSOs), primarily in Europe, but also in North America and Asia.[1] (See Table 1 and Figure 1.)

In most car-sharing organizations, participants have access to a fleet of cars parked at designated spots around town, often in a subscriber's neighborhood or at a major transit hub. Subscribers generally reserve a car in advance, although service on demand is possible if a car is available.[2] (In Switzerland, two thirds of surveyed members reported having a car within a 10-minute walk of their home, and 95 percent said they can get a car when they want one.)[3] Car sharers typically pay a refundable deposit, and sometimes a yearly membership fee.[4] They are also charged by the hour and by the kilometer for usage; these charges typically cover gas, insurance, and maintenance.[5]

*Links: pp. 38, 74, 152*

Car-sharing experiments can be traced back at least to the 1970s, but growth really took off in the late 1980s, when CSOs opened in Switzerland, and then in Germany.[6] These two nations now account for roughly 80 percent of the world's subscribers. Germany is home to 56,000 car sharers using nearly 2,400 cars, and Switzerland claims 43,000 members and more than 1,700 cars.[7]

In contrast to Europe, many cities in North America, especially in the United States, lack a good system of public transportation—a prerequisite for successful car-sharing. (Car-sharing is too expensive to use as a sole means of transport.) Still, car-sharing is growing rapidly in North America where conditions are favorable. One company operating in Boston and Washington, DC, saw membership grow more than sixfold between 2000 and 2001.[8] And older operations have proved successful in Montreal, Portland, San Francisco, Seattle, Toronto, and Vancouver.

Car-sharing carves out a niche for the automobile that plays to its greatest strength: as a highly flexible transportation option that allows travel to any destination or combination of destinations, on demand, with the capacity to carry cargo or passengers. It operates on the assumption that other transport needs—from daily commuting to trips to the corner market—are best met by other modes, such as mass transit, cycling, or walking. By using a mix of transportation, car sharers tap the automobile's assets while minimizing the expense, inconvenience, and extensive environmental toll of private ownership.

Owners of private cars have a strong economic incentive to drive rather than bus, cycle, or walk because of the heavy investment already made in their automobile and because the cost per trip of driving is relatively low.[9] But when car-sharing replaces private ownership, this incentive disappears, and people pay more attention to their cost per trip. Ironically, people who become car sharers often use the shared cars less and less over time as they become more aware of the relative costs of driving versus using mass transit, cycling, and walking.[10] Car sharers who previously did not own an automobile not surprisingly end up driving more, but they still travel less and use less fuel than a private owner.[11]

For people who can forgo a car for daily use, car-sharing can lower total transportation expenses. The American Automobile

## Table 1: Car-Sharing Members and Vehicles, by Region, 2001

| Region | Members | Vehicles |
|---|---|---|
| | (number) | |
| Europe | 112,701 | 4,865 |
| North America | 11,032 | 547 |
| Asia | 1,850 | 70 |
| Total | 125,583 | 5,482 |

*Source:* Worldwatch estimates based on discussions with various car-sharing organizations.

Association estimates that car ownership costs on average $7,600 per year in the United States for a car driven 24,000 kilometers (15,000 miles) annually.[12] People using a shared car for 16 hours a month, driving 128 kilometers (80 miles), would spend only $1,900 as members of a Washington, DC, CSO—a savings of $5,700 a year in driving expenses.[13] Even if they spend another $1,500 on mass transit, they can expect to save more than $4,000 annually.

Car-sharing is a good example of using services, rather than goods, to meet people's economic needs. In this case, subscribers rely on the service provided by a CSO rather than a private car to meet part of their transportation needs. This trend toward services reduces environmental impact by making the economy less materials-intensive. Indeed, each shared car is estimated to eliminate four cars—and all the rubber, metal, and glass that these represent—from the road.[14]

Because car sharers drive less than most car owners do, car-sharing brings real social and environmental benefits to cities. In Switzerland, people who give up their car when they join a CSO reduce car use by more than 70 percent a year, easing congestion and pollution.[15] And compared with nonmembers, subscribers to the Mobility CSO in Switzerland use about 55 percent less fuel annually.[16] Yet they are not socially isolated: they end up travelling about 10 percent more a year than before they started car-sharing—via mass transit, bicycle, motorbike, or walking.[17] And because shared cars are used more intensively than privately owned vehicles (which typically sit idle most of the time), sharing reduces the need for parking spaces.

Some European car-sharing agencies have begun to link car-sharing with other transportation options and with other car-sharing organizations. In 1998, Swiss Federal Railways and Mobility CarSharing Switzerland introduced a combined season pass that allowed access to shared cars and to trains throughout

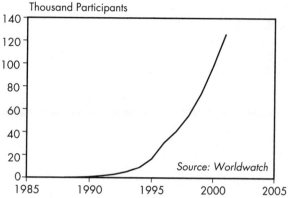

Thousand Participants

**Figure 1: Growth in Car-Sharing Worldwide, 1988–2001**

the country.[18] And European Car-Sharing, a consortium of these organizations, makes cars from its members available to subscribers in more than 80 cities across the continent.[19]

The future of car-sharing appears to be bright. In Switzerland, some 600,000 people— 9 percent of the population, and 15 times the current car-sharing subscription base—are estimated to be interested in signing up.[20] Meanwhile, rental companies are jumping into the game. In May 2000, San Francisco's rapid transit authority and Hertz jointly launched a commercial "station car" rental program.[21]

# Sprawling Cities Have Global Effects   *Molly O. Sheehan*

Cars and highways are stretching cities to new limits, as cars require more space than other forms of urban transportation do. A lane of light rail, for example, can move four to eight times more people per hour than a lane of highway can, while 10–20 bicycles can be parked in the space needed for one car.[1]

Although there is no single global measure of car-dependent urban development, or "sprawl," census data do reveal more spread-out cities in some parts of the world— most dramatically in the United States.[2] One analysis of U.S. census statistics found that the number of people living in 58 U.S. metropolitan areas rose 80 percent between 1950 and 1990, while the land covered by those areas expanded 305 percent.[3] Another study found that even in 11 urbanized areas in the United States where population decreased between 1970 and 1990, the amount of land covered by those urbanized areas increased.[4]

Links: pp. 52, 74, 88, 150

Researchers trying to track urbanization patterns worldwide are thwarted by inaccurate or outdated census data, so satellite remote sensing has emerged as an important tool. For example, Landsat images of Shenzhen in China reveal a rapid increase in built-up area: 25 percent growth between 1992 and 1996 alone.[5] One set of studies undertaken in the United States in the 1990s combined Landsat images with historical maps and census data to show land growth outpacing population growth in the San Francisco– Sacramento and Washington, DC–Baltimore regions.[6] Scientists plan to use this approach for other cities.[7]

Many cities are located on prime agricultural sites, so urban expansion paves over valuable farmland. Researchers have used images from a U.S. satellite to create a map of nighttime city lights that corresponds well to census estimates of urban area.[8] When researchers compared the area covered by cities to the U.N. Food and Agriculture Organization's digital soil map, they found that although only 3 percent of the U.S. land surface is urbanized, the best soils are being developed first.[9] Suburban roads and houses supplant more than 1 million hectares of farmland each year in the United States.[10] In China, the government estimates that some 200,000 hectares of arable land disappear each year under city streets and developments.[11]

Sprawl is also linked to global climate change. Carbon dioxide, released in large quantities by fossil fuel combustion, is one of the most important heat-trapping "greenhouse gases" warming the atmosphere.[12] Between 1990 and 1998, road transportation was the fastest growing source of carbon emissions from fuel burning.[13] (See Figure 1.) Researchers studying transportation and land use in cities worldwide find higher carbon emissions per person in less densely populated, sprawling urban areas.[14] (See Figure 2.)

If average temperatures continue to rise as scientists project, the consequences are likely to include sea level rise, deadly heat waves, and an expanded range for disease vectors—and most cities will be vulnerable to one or more of these threats.[15] Many cities are located along coasts, which are most endangered by sea level rise. Indeed, two thirds of the world's population lives within 60 kilometers of a coast, and this number is expected to rise to three fourths by 2010.[16] Even a small hike in average temperature can increase the risk of heat waves, which can be especially deadly in cities because

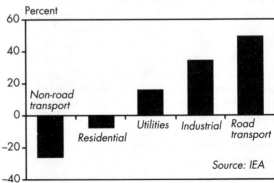

Figure 1: Change in World Carbon Emissions from Fossil Fuel Burning, by Sector, 1990–98

hot weather accelerates the chemical reactions that produce urban smog.[17]

Perhaps the most important global implication of sprawl, however, is that more of the world's people are experiencing the local health effects of car-centered development, including air pollution and road deaths. Studies in Europe show that in some countries pollution from motor vehicles actually kills more people than vehicle accidents do.[18] The death toll from vehicle accidents alone is not insignificant: nearly a million people—mostly pedestrians—are killed on the world's roads each year.[19]

Motor vehicles also harm human well-being by impeding other forms of traffic and causing costly delays. Cities such as Atlanta, Bangkok, and Jakarta are less densely populated than Paris, Moscow, or Shanghai, yet suffer worse traffic delays because they have neither effective public transit systems nor adequate facilities for bicyclists and pedestrians.[20] Every day Atlanta loses more than $6 million to traffic delays, and Bangkok more than $4 million.[21] But such estimates only value hours that could have been spent working; it is harder to measure the loss to society of time that could have been used to care for children or build friendships in a community.

Nearly half the world—2.8 billion people—lived in urban agglomerations in 1999, almost four times as many as in 1950.[22] And that number is projected to rise significantly. The way cities are built directly affects the lives of many more people today than in years past, and will likely continue to do so in the decades to come.

Although the situation varies from country to country, economic forces tend to favor sprawl: the price of construction falls with distance from city centers, the up-front price of building a road network is far less than investing in a public transit system, and the price of any given car trip in many cities is less than that of bus or rail travel.[23]

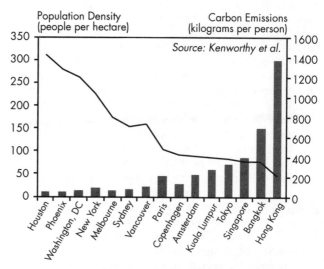

**Figure 2: Population Density Versus Carbon Emissions from Transportation, 1990**

These prices, however, do not reflect the costs to society of car-dependent cities. Various researchers have estimated the costs of road transport not covered by drivers—including air pollution, noise, traffic delay, road damage, and accidents—to be around 5 percent of gross domestic product in industrial countries such as the United States, and even higher in some developing-country cities.[24]

Governments could alter price signals that favor sprawl by minimizing their spending on new roads, sewers, and other infrastructure in outlying areas, by removing barriers to investment in central locations, and by adjusting the taxes of various transportation modes. Bicycling imposes few costs to society, so in countries where a bicycle constitutes a serious investment, governments could slash the luxury tax on bikes and help lower the barriers to purchase by underwriting loans.[25] Governments could also increase the price of driving by raising motor fuel taxes, charging heavy trucks for the extra wear they impose on roads, introducing fees for driving on congested roads at peak travel times, and substituting transit tickets for parking privileges.[26]

# Teacher Shortages Hit Hard

*Kathleen Huvane*

The world will require more than 18 million additional teachers in the coming decade if it is to reach universal primary education goals by 2015.[1] Teacher shortages persist in a variety of forms and regions (see Figure 1) that have disparate impacts—affecting access, duration, and quality of schooling for children.[2]

Generalizations sometimes mask inequities within nations as well. The richest 10 percent of El Salvador's youth, for instance, receive 8.6 more years of schooling than the poorest 10 percent, indicating an acute lack of teachers for the most needy.[3] Persistent shortages plague remote rural and low-income urban areas of the United States, where turnover rates are highest and the greatest percentage of teachers are uncertified, especially in schools with a high proportion of non-Caucasian students.[4] Seventy percent of American schools with "low minority" populations have math teachers with a degree in math, compared with just 42 percent in "high minority" schools.[5]

Links: pp. 88, 90, 148

Mortality from AIDS has reduced populations of both students and teachers throughout sub-Saharan Africa.[6] By 2010, for example, Zimbabwe's primary school population will likely decline by 24 percent.[7] And eight teachers die every week in Côte d'Ivoire from AIDS-related illness.[8] All employment sectors have been affected by AIDS, and many teachers have left the profession to fill gaps left in more lucrative professions.[9]

The dearth of trained teaching professionals reinforces the cycle of under-enrollment. One third of all children in developing countries attend school for less than five years.[10] Although student enrollment has increased in every region of the world since 1990, more than one quarter of the children in South Asia and 40 percent of all children in Africa did not have access to a formal education in 1998.[11]

Many highly qualified teachers from developing countries are being recruited to fill positions in U.S. and European schools.[12] In 2001, teachers from such countries as Barbados and Jamaica were recruited by New York City public schools to address an 8,000-person teacher shortage.[13] Lured by salaries up to four times higher than in their home countries, these teachers are often some of the most educated and fluent English speakers in their fields.

Salaries, the single largest expenditure for schools, determine the number of teachers that can be hired. As public resources diminish, parents and other supporters have to pay more for schooling. Some 40 percent of education funding in Chile, Peru, and the Philippines comes from private sources, and Oxfam estimates that the poorest 40 percent of developing-country families will spend one tenth of their income to send two children to primary school.[14]

Education spending is often curtailed in countries undergoing economic transition or crisis. Economic reforms in Zambia outlined by the International Monetary Fund in 1991 led to a 25-percent reduction in education spending in just three years.[15] Private-sector growth in education systems is encouraged in developing countries like Côte d'Ivoire, where 60 percent

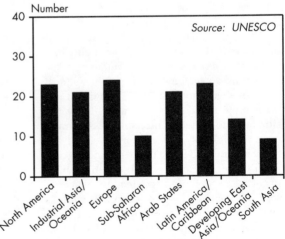

Figure 1: **Teachers per Thousand Population, by Region, 1997**

of secondary schools have been privatized—a trend that can exacerbate inequities as the poorest, especially girls, are excluded.[16]

Teacher shortages in industrial countries are exacerbated during times of economic growth because existing and potential teachers are drawn to higher-paying jobs. Ironically, recent downturns in the global economy may have a positive impact on the teacher supply: in California's Silicon Valley, shortages have diminished since the technology economy crashed in 2000.[17]

When scarcities intensify, teachers are assigned to positions for which they are not adequately trained. Half of California's new schoolteachers in 2000, for example, had either no credentials or were inadequately prepared for the subjects they taught.[18] Sub-par test scores in Germany have been partly attributed to low teacher morale and shortages in certain subject areas.[19]

Training and supporting new teachers can be costly, and returns from these investments are often never realized. For instance, half of all new teachers in city schools in the United States quit within five years.[20] Nearly half of Australia's teacher population is over 45, and the number of university students in education programs there fell by 33 percent between 1991 and 1998.[21] A third of Germany's teachers reportedly feel overworked, and two thirds retire early.[22]

Children in industrial countries spend more time in school and benefit from smaller class sizes. In contrast, compulsory education in Colombia and Nepal lasts just five years—half as long as in France and Australia.[23] Student-teacher ratios vary widely across the world, from a low of 1:8 in Libya to a high of 1:72 in Mali.[24] The typical primary school teacher in the Democratic Republic of Congo in 1995 had 24 more pupils in his or her class than a teacher in Spain, a gap indicative of regional trends.[25]

A shortfall of qualified teachers in math, science, special education, and bilingual education afflicts schools in the United States and many European countries.[26] Such prob-

lems also exist in developing countries, where less than 1 percent of children with special needs attend school, but they are often overshadowed by more pressing educational crises.[27] Some governments are under pressure to find enough teachers to meet the demands of demographic "youth bulges." Half the population of Egypt, Iran, Iraq, Saudi Arabia, and Syria is under 25, while over 60 percent of Pakistan and Afghanistan's populations fall into that category.[28]

Teacher shortages are also a result of national disasters and conflict. Some 80 percent of children not enrolled in school live in crisis or post-crisis countries, where the teaching staff who remain are commonly undertrained and underpaid.[29] Teachers have even been specifically targeted for political killings by regimes such as the Khmer Rouge, which decimated Cambodia's teacher population in the 1970s—a loss from which it has yet to recover.[30]

A lack of teachers with diverse cultural and ethnic backgrounds constitutes another type of shortage, as students are deprived of role models and perspectives from their own communities. Over a third of all U.S. students are minorities, but 87 percent of the teachers are Caucasian.[31] Teacher shortages can also be gender-based. Countries with low female literacy and enrollment rates are also likely to have lower percentages of female teachers.[32] Though women dominate the teaching profession globally, they are rare in remote rural areas of developing countries.[33]

Free and compulsory education is identified as a fundamental human right by the United Nations.[34] It can elevate living standards for individuals and societies as a whole, with positive impacts measured across a broad range of economic and social indicators, including increased gender equity, improved health, higher incomes, and lower levels of population growth. While effective education can take many forms, there is little dispute about the fact that a skilled teaching force is central to the actualization of this commitment.

# Women Subject to Violence
*Danielle Nierenberg*

One in three women worldwide has experienced abuse in her lifetime.[1] And in some nations, according to the World Health Organization, the number of women ever abused ranges from 16 to 52 percent.[2] Abuse from an intimate partner—the most common form of violence against women—occurs in all countries, transcending economic, cultural, and religious boundaries.[3] But these numbers are only estimates at best—the shame, fear, and lack of legal rights that accompany gender inequality keep many women from reporting their attackers or even acknowledging that abuse is a problem throughout their lives.

Abuse shadows women from birth—and even before.[4] (See Table 1.) Sex-selective abortions, female infanticide, and neglect of girl children are common in India, China, and other nations.

Links: pp. 88, 94, 144, 148

UNICEF estimates that more than 60 million girls worldwide are considered "missing" because they were aborted, killed shortly after birth, or hidden from authorities.[5] In the Indian state of Haryana, the sex ratio has increasingly favored males since the early 1990s—160 males are born for every 100 females.[6] China's most recent census shows that 117 boys are born for every 100 girls, and in the most remote regions the difference is even higher.[7] By comparison, in Germany usually 96 boys are born for every 100 girls.[8]

As children and adolescents, girls experience such familial, educational, and cultural abuses as enforced malnutrition, incest, female genital mutilation (FGM), denial of medical care, early marriage, prostitution, and forced labor.[9] An estimated 140 million women and girls have undergone debilitating mutilation of their genitals, and another 2 million are at risk of being subjected to this practice each year because of their ethnic or religious backgrounds.[10] FGM causes its victims to suffer a lifetime of painful urination, menstruation, and sexual intercourse, as well as difficulties during childbirth. It also increases women's vulnerability to HIV/AIDS and other sexually transmitted diseases due to the use and reuse of unsanitary instruments.[11]

According to the U.S. Department of Justice, young women—teenagers to women in their mid-twenties—are nearly three time as vulnerable as older women to attack by a husband, boyfriend, or former partner.[12] But older victims of domestic violence—those between 35 and 49 years of age—are more likely than younger ones to be killed.[13] Overall, women in the United States are 60 percent more likely than men to be killed by an intimate partner.[14]

Violence against women often includes more than physical or verbal assaults. A third to half of physically abused women also report forced sex.[15] In fact, a study in Leon, Nicaragua, found that only 2.5 percent of women abused by their partners had *not* been sexually assaulted.[16] Used as a weapon, sexual violence in all its forms—coerced sex, rape, incest—inhibits women's ability to control their own reproductive health. Women in conflict situations are particularly vulnerable when rape is used as tool of intimidation. During the war in the Balkans, between 20,000 and 50,000 women and girls were raped in Bosnia-Herzegovina.[17]

Children, and especially girls, suffer from domestic violence as well. According to UNICEF, almost two thirds of the children who live in families where the mother is abused by a husband or boyfriend are also beaten.[18] Children of women who are beaten by their domestic partners are more likely than other children to die before the age of 5.[19]

"Too often, women and girls cannot say no to unwanted and unprotected sex without fear of reprisal," notes Noeleen Heyzer, Executive Director of UNIFEM, the U.N. Development Fund for Women.[20] Many men consider sex their unconditional right, and fear often prevents women from discussing contraceptives or their sexual rights with partners. UNICEF reports that women in Kenya and Zimbabwe hide their birth control pills in fear that their husbands will discover that "they no longer control their wives' fertility."[21] Forced sexual initiation also increases the risk of HIV infection among girls and women. More than two thirds of girls in South Africa do not choose to have sex the first time but are instead coerced

## Table 1: Selected Examples of Violence Against Women

| | |
|---|---|
| Female Infanticide | In Punjab, India, only 793 girls were born for every 1,000 boys. |
| Female Genital Mutilation | In Ethiopia, nearly 85 percent of girls have undergone mutilation of their genitalia; in Somalia, the figure is more than 95 percent. Worldwide, more than 6,000 girls per day are in danger of undergoing these procedures due to their ethnic or religious backgrounds. |
| Rape | In the United States, 1.5 million women are raped annually and 14–20 percent of women will be raped in their lifetimes. |
| Murders | In India, more than 5,000 brides are killed annually because their families are unable or unwilling to pay the dowry promised at marriage. |
| Honor Killings | As many as 5,000 young women died at the hands of their parents or other relatives in 2000 for "shaming" their families by having sex, socializing with boys, or becoming victims of rape. |
| Suicides | In China, suicide is the leading cause of death for women between the ages of 20 and 34. |

Source: See endnote 4.

and raped, increasing their chances of contracting AIDS in a nation where 10 percent of the population is HIV-positive.[22]

Poverty can exacerbate abuse, exploitation, and violence against women. An estimated 4 million women and girls are bought and sold worldwide each year.[23] Traffickers target economically depressed families, promising to find work and schooling for their daughters in the city. Most of these young women and girls then become prostitutes: at least 10,000 enter the commercial sex trade in Thailand each year, and roughly 7,000 Nepali girls are brought into India annually for prostitution.[24]

Ironically, increasing women's economic participation can contribute to a sense of inadequacy among men, leading them to use violence as a means of "control." In addition to being beaten, maquiladora workers in Mexico report being deprived of their earnings by their husbands, as do women in microcredit programs in parts of Peru and Bangladesh.[25] In Papua New Guinea, the main reason female teachers gave for turning down promotions was "the fear that it would provoke their husbands to more violence."[26]

In extreme cases, women may try to end

their lives to escape abuse. According to Radhika Coomaraswamy, the U.N. envoy on violence against women, "the suicide rate among women is high in conservative and repressive societies."[27] Under restrictive religious regimes, "some women see suicide as the only way out."[28] Females in some parts of Turkey, Afghanistan, and Iran account for as much as 80 percent of all suicides.[29] Nearly 500 women a day kill themselves in China alone.[30]

In 1995, the Fourth World Conference on Women in Beijing gave priority to eradicating violence against women—calling it an "obstacle to the achievement of the objectives of equality, development and peace."[31] An end to the violence requires an end to the discrimination that women face in every aspect of their lives—from inequities in education and the workplace to the lack of control they have over their sexual and reproductive lives.

# Voter Participation Declines

*Molly O. Sheehan*

Voter participation in competitive elections worldwide declined slightly between 1945 and 1979, from an average of 78 percent to 76 percent, and then rose to 79 percent in the 1980s, according to the Stockholm-based International Institute for Democracy and Electoral Assistance (IDEA).[1] But overall turnout at elections dipped to 71 percent in the 1990s.[2] (See Figure 1.)

Several factors help to explain the drop-off. Since the 1970s, voter participation in many established democracies has declined gradually.[3] Many nations of the developing world that have introduced elections since the 1960s have young populations, and youth in all nations tend to vote less than their elders do.[4] Also, there was a sharp rise in the total number of elections in the 1990s, which was spurred by the newly democratizing nations of the former Soviet bloc. These countries registered relatively low voter turnout on average, although their participation levels are increasing with time.[5] Whereas there were some 294 presidential and parliamentary elections in the 1980s, there were 603 such elections in the 1990s.[6]

Despite the dip, voters have turned out in force in a wide range of nations all across the world since 1945.[7] (See Table 1.) The elections with the best turnouts since 2000 similarly have occurred in a cross-section of nations: more than 90 percent of eligible voters participated in parliamentary elections in Ethiopia, Tajikistan, and Guyana in 2000 and in presidential elections in Chile in 2000 and Seychelles in 2001.[8] Five of the top democracies in voter turnout—Australia, Belgium, Liechtenstein, Nauru, and Singapore—enforce compulsory voting laws, which could help explain their high participation rates.[9] There is no significant correlation between a nation's voter turnout and its wealth, the number of years it has been a democracy, or its literacy level.[10]

Although the existence of elections does not guarantee a high quality of life, the 86 electoral democracies that receive the highest ranking of "free" by the independent monitoring group Freedom House are also the most prosperous.[11] This organization finds that the share of world population living in "free" democratic states grew from 36 percent in 1981 to 41 percent in 2001.[12] The worst nations in this ranking are Afghanistan, Cuba, Equatorial Guinea, Iraq, Libya, Myanmar (formerly Burma), North Korea, Saudi Arabia, Sudan, Syria, and Turkmenistan.[13]

Despite the promise of a host of first-ever democratic elections in Eastern Europe and Central Asia in the 1990s, several nations in these regions saw progress toward free and fair elections erode. The March 2000 presidential elections in Russia suffered from heavy government pressure on the press, voter fraud marred the November 2000 elections in Azerbaijan, and the government of the Ukraine has increasingly interfered in press coverage of public affairs.[14]

Today, 21 of 53 nations in Africa are electoral democracies, more than ever before, but corruption continues to mar many campaigns and elections in this region.[15] Reports of political killings overshadowed Zimbabwe's 2002 presidential campaign, as the government led by President Robert Mugabe cracked down on the media and intimidated opponents.[16] Despair over entrenched government graft

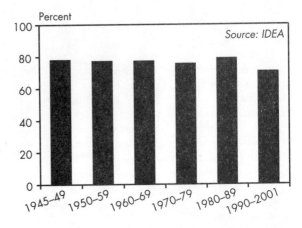

Figure 1: Voter Turnout in Elections, 1945–2001

## Table 1: Top 15 Nations in Voter Participation, 1945–2001

| Country | Elections (number) | Average Share of Voters Participating in National Elections (percent) |
|---------|------|-------|
| Australia | 22 | 94.5 |
| Singapore | 8 | 93.5 |
| Uzbekistan | 3 | 93.5 |
| Liechtenstein | 17 | 92.8 |
| Belgium | 18 | 92.5 |
| Nauru | 5 | 92.4 |
| Bahamas | 6 | 91.9 |
| Indonesia | 7 | 91.5 |
| Burundi | 1 | 91.4 |
| Austria | 17 | 91.3 |
| Angola | 1 | 91.2 |
| Mongolia | 4 | 91.1 |
| New Zealand | 19 | 90.8 |
| Cambodia | 2 | 90.3 |
| Italy | 15 | 89.8 |

Source: See endnote 7.

appeared to be at least one of the factors contributing to violence in Kenya's slums in the run-up to that nation's 2002 election.[17]

On a more positive note, a number of elections since 2000 have served to further peace and human rights. In Ghana, a president who seized power in a 1981 coup and ruled for nearly two decades was prevented from continuing for a third term by the nation's constitution. The nation held its first-ever democratic elections in 2000, in which more than 60 percent of the voting population turned out to elect opposition leader John Kufour as president.[18]

After years of bloodshed in Yugoslavia under Slobodan Milosevic, the free election of Vojislav Kostunica in 2000 heralded a new era.[19] Similarly, in the province of Kosovo, the first-ever democratic elections were held in 2000.[20]

In Mexico, the election of Vicente Fox in 2000 ended more than 70 years of virtual one-party government.[21] The resignation of Alberto Fujimori in Peru amidst a corruption scandal was followed by gains in political freedom and the fair election of Alejandro Toledo in 2001.[22]

Some 86 percent of voters turned out to participate in the first democratic election in East Timor in 2001.[23] The elections created a new parliament two years after East Timor voted to affirm its independence from Indonesia.[24] And in February 2001, some 90 percent of registered voters in Bahrain turned out to support a referendum to establish a democratically elected chamber in parliament and to set up an independent judiciary.[25]

Some more established democracies held elections with record low turnouts in recent years. The U.S. presidential election of 2000 attracted only 51 percent of eligible voters, although it received worldwide attention as a result of the debate over the counting of ballots in the closely contested race.[26] The United Kingdom also had a record low turnout for its parliamentary elections in 2001, with only 59 percent of eligible voters participating.[27]

In countries that hold competitive elections, advocates of environmental protection and social justice have the opportunity to raise the profile of these issues at the ballot box. In Western Europe, strong, pro-environment "green" political parties have pushed environmental issues onto the public agenda in many nations.[28] And as urban sprawl has emerged as a growing concern in the United States, voters approved hundreds of local and state ballot initiatives to address the problem in the 1998 and 2000 elections.[29]

# Military Features

Progress Against Landmines

# Progress Against Landmines
*Michael Renner*

Anti-personnel landmines have taken a heavy toll for decades.[1] Not only do they kill and maim indiscriminately, they make fertile land unusable, inhibit travel, prevent farmers from getting their produce to markets, discourage repatriation of refugees, and hinder reconstruction efforts after wars end.

Since the passage of the international mine ban treaty in 1999, however, significant headway has been made in recent years in battling this problem.[2] A growing number of governments are joining the treaty. And the International Campaign to Ban Landmines (ICBL), an international coalition of nongovernmental organizations, concludes that there is reduced use of anti-personnel mines, a dramatic drop in their production, a near-complete halt to exports, destruction of stockpiled mines at a rapid pace, growing amounts of land that has been cleared of mines, and fewer victims in affected countries.[3]

Link: p. 94

The Mine Ban Treaty was hammered out in just 14 months, opened for signature in December 1997, and entered into force in March 1999—lightning speed compared with the usual process of international negotiating and treaty-making.[4] Its adoption capped a highly successful campaign by the ICBL, which started in 1992.[5]

As of October 2001, 142 countries had signed or ratified the treaty.[6] Almost all African countries, heavily affected by mines, have signed on. In the western hemisphere, only the United States and Cuba have not joined in; in the European Union, only Finland has not. Most of the Middle East and many Asian nations have so far declined to join the treaty. Unfortunately, the 53 holdouts around the world include three of the five permanent members of the Security Council (the United States, Russia, and China) and some other major producers, including India and Pakistan.[7]

In its 2001 *Landmine Monitor* report, the ICBL puts the number of countries still producing anti-personnel mines at 14, among them Russia, Egypt, Iran, Iraq, China, India, Pakistan, North Korea, and South Korea.[8] The United States is included in this category as well; although it has not manufactured any mines since 1996, it has refused to adopt an official moratorium or ban.[9] Encouragingly, 41 nations have now ended production, including 8 of the 12 major past producers, all of them in Europe.[10]

The ICBL estimates that about 230–245 million anti-personnel mines remain stockpiled in 100 countries.[11] Non-signatories to the treaty hold the vast majority—some 215–225 million.[12] China alone is believed to have 110 million mines, followed by Russia with 60–70 million, and the United States with 11 million.[13] Ukraine, Pakistan, India, and Belarus each hold about 4–6 million.[14] The treaty requires that stockpiles be eliminated, and some 27 million anti-personnel mines have been destroyed in recent years in as many as 50 countries.[15] By late 2001, 28 countries had destroyed their arsenals, and another 19 are well along in this task.[16]

The U.S. State Department estimates that there are 45–50 million landmines buried in nearly 60 countries.[17] But *Landmine Monitor 2001* reports that 90 countries and 11 non-sovereign territories (such as Chechnya, Iraqi Kurdistan, Palestine, and Somaliland) are affected.[18] Only about one third of them have undertaken surveys or undergone a systematic assessment.[19] During 2000 and early 2001, mine clearance operations were carried out in 76 countries and territories.[20]

No one knows how many people fall victim to mines each year. The tally from reported incidents in 2000 was somewhat less than 10,000 casualties.[21] But a significant number of incidents are believed to go unreported. The ICBL estimates that mine explosions kill or maim 15,000–20,000 persons annually.[22] While this is still a very large number, it is down from earlier estimates of 26,000–30,000 casualties.[23] Landmines continue to be a danger long after a conflict comes to an end; most casualties occurred in countries no longer at war.[24]

More than $1 billion has been made available for demining activities during the past decade.[25] Though tracking available money is difficult, the ICBL puts spending in 2000 at $224 million.[26] The amounts available have increased, but the United Nations judges available resources still too limited to meet the needs of affected countries.[27] The leading funders are the United States, Norway, the United Kingdom, Sweden, Germany, and Japan.[28] By the end of 2001, the United States had provided more than $500 million in demining assistance to 40 countries.[29] The bulk of funds went to Afghanistan, Angola, Bosnia, Cambodia, and Mozambique.[30]

Afghanistan has been heavily mined since the late 1970s. The upsurge in fighting that accompanied the U.S. air campaign against the Taliban regime and Al Qaeda forces in late 2001 added unknown quantities of unexploded ammunitions to the demining challenge.[31] Just how many mines are scattered was unknown even before the most recent turn of events. The U.S. State Department estimates them at 4 million (see Table 1), down from its 1998 estimate of 10 million.[32] Some observers insist the number is 1 million or less, whereas the U.S. Campaign to Ban Landmines uses an estimate of 8–10 million.[33]

In Afghanistan, some 723 square kilometers have been found to be mine-infested, but additional mined areas are being detected at the rate of 12–14 square kilometers a year.[34] Mines have severely reduced the amount of Afghan agricultural and grazing land safely accessible. The U.N. Development Programme and the World Bank estimate that at current funding levels, it will take 7–10 years to clear roughly half of the contaminated areas, some 344 square kilometers of the most productive land, which would allow most Afghans to resume a more normal life.[35]

In 2000, there were more than 1,000 recorded mine casualties in Afghanistan, down from more than 7,200 in 1993.[36] But the real number could easily be 50–100 percent larger.[37] Indeed, a State Department estimate uses a figure of 2,400.[38] And the Organization for Mine Clearance and Afghan Rehabilitation, a U.N.-sponsored agency, believes annual casualties to be as high as 4,000.[39]

The populations of many mine-affected countries will likely confront the dangers and uncertainties of mines for decades, if not centuries. A recent assessment of Cambodia's situation concluded that landmine clearance may take 200–300 years at current removal rates.[40]

## Table 1: Estimated Effect of Anti-Personnel Landmines, Selected Countries, 2000

| Country | Number of Landmines | Land Area Affected (square kilometers) | Share of Territory (percent) | Landmine Victims, 2000 |
|---|---|---|---|---|
| Afghanistan | 4 million | 723 | 0.1 | 2,400 |
| Angola | 200,000–6 million | 634,547 | 50.9 | 840 |
| Armenia | 100,000 | 2,500 | 8.0 | 8 |
| Bosnia-Herzegovina | 1 million | 4,200 | 8.2 | 87 |
| Cambodia | 300,00–1 million | 2,000 | 1.1 | 811 |
| Colombia | 70,000 | 248,216 | 21.8 | n.a. |
| Croatia | 1–1.2 million | 4,000 | 7.1 | 22 |
| Egypt | 5–7.5 million | 2,800 | 0.2 | n.a. |
| Ethiopia | 1.5–2 million | 2,000 | 0.2 | 15 |
| Viet Nam | 3.5 million | n.a. | n.a. | 2,000 |

*Source*: U.S. Department of State, *To Walk the Earth in Safety: The United States Commitment to Humanitarian Demining* (Washington, DC: November 2001); Worldwatch calculations.

# Notes

## AQUACULTURE PRODUCTION INTENSIFIES (pages 24–25)

1. Data for 1984 from U.N. Food and Agriculture Organization (FAO), *Aquaculture Production Statistics, 1984–93* (Rome: 1993); 1999 data from idem, *Fishery Statistics: Aquaculture Production* (Rome: 2001), p. 47.
2. Adele Crispoldi, U.N. Fisheries Department, Fishery Information, Data, and Statistics Service, e-mail to author, 21 January 2002.
3. "Bangkok Declaration and Strategy for Aquaculture Development Beyond 2000," in Rohana P. Subasinghe et al., eds., *Aquaculture in the Third Millennium: Technical Proceedings of the Conference on Aquaculture* (Bangkok: Network of Aquaculture Centres in Asia-Pacific and FAO, 2000), p. 463.
4. "Chapter 7: Fisheries," in FAO, *Agriculture: Toward 2015/30*, technical interim report (Rome: 2001), p. 172.
5. FAO, *Fishery Statistics*, op. cit. note 1, p. 47.
6. "Chapter 7," op. cit. note 4, p. 173.
7. FAO, *Fishery Statistics*, op. cit. note 1, p. 46.
8. Ibid., p. 47.
9. Crispoldi, op. cit. note 2; Reg Watson and Daniel Pauly, "Systematic Distortions in World Fisheries Catch Trends," *Nature*, 29 November 2001, pp. 534–36.
10. FAO, *Fishery Statistics*, op. cit. note 1, p. 47.
11. Ibid.
12. Ibid.
13. Ibid., pp. 47, 75–76.
14. "Chapter 7," op. cit. note 4, p. 172.
15. FAO, *Fishery Statistics*, op. cit. note 1, p. 56.
16. Ibid.
17. FAO, Committee on Fisheries (COFI), "Report of the Seventh Session of the Sub-Committee on Fish Trade, Bremen, Germany, 22–25 March 2000," *FAO Fisheries Report. No. 621* (Rome: 25 March 2000).
18. COFI, "Item 6 of the Provisional Agenda: Globalization and Implications for International Fish Trade and Food Security," Meeting Document Prepared for the Seventh Session of the Sub-Committee on Fish Trade, Bremen, Germany, 22–25 March 2000.
19. FAO, Economic and Social Department, Commodities and Trade Division, *Fisheries Products: Commodity Notes* (Rome: November 2000).
20. Rosamond L. Naylor et al., "Effect of Aquaculture on World Fish Supplies," *Nature*, 29 June 2000, pp. 1017–24; John Tibbetts, "Aquaculture: Satisfying the Global Appetite," *Environmental Health Perspectives*, July 2001, pp. A321, A323.
21. James H. Tidwell and Geoff L. Allan, "Fish as Food: Aquaculture's Contribution—Ecological and Economic Impacts and Contributions of Fish Farming and Capture Fisheries," *EMBO (European Molecular Biology Organization) Reports*, vol. 2, no. 11 (2001), pp. 958–63.
22. Ibid.
23. FAO, *Fishery Statistics*, op. cit. note 1, p. 56.
24. Tibbetts, op. cit. note 20, p. A323.

## GRAIN HARVEST LAGGING BEHIND DEMAND (pages 26–27)

1. U.S. Department of Agriculture (USDA), *Production, Supply, and Distribution*, electronic database, Washington, DC, updated December 2001.
2. Ibid.
3. Ibid.; population data from U.S. Bureau of the Census, *International Data Base*, electronic database, Suitland, MD, updated 10 May 2000.
4. USDA, op. cit. note 1.
5. Ibid.; USDA, Foreign Agricultural Service, *Grain: World Markets and Trade* (Washington, DC: December 2001).
6. USDA, op. cit. note 1; U.N. Food and Agriculture

Organization, *The State of Food Insecurity in the World 2000* (Rome: 2000), p. 4.

7. USDA, op. cit. note 1.

8. Ibid.

9. Ibid.

10. Lester R. Brown, "Worsening Water Shortages Threaten China's Food Security," *Eco-Economy Update 1* (Earth Policy Institute, Washington, DC), 4 October 2001.

11. Lester R. Brown and Brian Halweil, "China's Water Shortage Could Shake World Food Security," *World Watch*, July/August 1998, p. 18.

12. Hsin-Hui Hsu, Bryan Lohmar, and Fred Gale, "Surplus Wheat Production Brings Emphasis on Quality," in USDA, Economic Research Service, *China: Agriculture in Transition* (Washington, DC: November 2001), pp. 17–25.

13. Ibid.

14. USDA, op. cit. note 1.

15. Ibid.

16. Ibid.

17. Ibid.

18. Ibid.

19. Ibid.

20. Ibid.

21. Ibid.

## MEAT PRODUCTION HITS ANOTHER HIGH (pages 28–29)

1. U.N. Food and Agriculture Organization (FAO), *FAOSTAT Statistics Database*, at <apps.fao.org>, updated 7 November 2001.

2. Ibid.

3. Ibid.

4. Ibid.

5. Ibid.

6. Ibid.

7. Ibid.

8. U.S. Department of Agriculture (USDA), Foreign Agricultural Service, *Livestock and Poultry: World Markets and Trade* (Washington, DC: October 2001).

9. Ibid.

10. Ibid.

11. FAO, op. cit. note 1.

12. USDA, op. cit. note 8.

13. Ibid.

14. Ibid.

15. FAO, op. cit. note 1.

16. Ibid.

17. USDA, op. cit. note 8.

18. Ibid.

19. Ibid.

20. Ibid.

21. Ibid.

22. Ibid.

23. Ibid.

24. Ibid.

25. Ibid.

26. Ibid.; USDA, Foreign Agricultural Service, *Grain: World Markets and Trade* (Washington, DC: December 2001).

27. USDA, op. cit. note 8.

28. Ibid.

29. Ibid.

## COCOA PRODUCTION JUMPS
(pages 30–31)

1. U.N. Food and Agriculture Organization (FAO), *FAOSTAT Statistical Database*, at <apps.fao.org>, updated 7 November 2001.

2. Ibid.

3. Ann Gray, LMC International, "The World Cocoa Market Outlook," May 22, 2001, at <www.acri-cocoa.org/acri/LMCrep1.pdf>, viewed 18 December 2001.

4. FAO, op. cit. note 1.

5. Ibid.; Robert A. Rice and Russell Greenberg, "Cacao Cultivation and the Conservation of Biological Diversity," *Ambio*, May 2000, pp. 167–73.

6. World Bank, *Country-At-A-Glance Data Profiles for Côte d'Ivoire and Ghana*, at <www.worldbank.org>, viewed 4 January 2002.

7. The Ministry of Primary Industries Malaysia, *Cocoa Report*, at <www.kpu.gov.my/commodity/cocoa1.htm>, viewed 15 December 2001; FAO, Economic and Social Department, Commodities and Trade Division, updated February 2001, at <www.fao.org/waicent/faoinfo/economic/ESC/esce/cmr/cmrnotes/CMRcoce.htm>, viewed 2 January 2002.

8. FAO, op. cit. note 7.

9. FAO, op. cit. note 1.

10. Heather White, "Just Desserts," at <www.lohasjournal.com/so00/desserts.html>, 2001, viewed 12 December 2001.

11. David Pacchioli, "Cocoa Futures," *Penn State Online Research*, May 2001, at <www.rps.psu.

edu/0105/cocoa.html>, updated 4 December 2001; Center for New Crops and Plants Products, Purdue University, "*Theobroma cacao L.*," from James Duke, "Handbook of Energy Crops," at <www.hort.purdue.edu/newcrop/duke_energy/Theobroma_cacao.html#Yields%20and%20Economics>, viewed 11 December 2001.

12. Rice and Greenberg, op. cit. note 5, pp. 167–73.

13. International Cocoa Organization (ICCO), at <www.icco.org>, viewed 17 December 2001.

14. Dixxon Chok, "Cocoa Development & Its Environmental Dilemma," Shade-grown Cacao Integrated Research Center, Smithsonian Migratory Bird Center, at <natzoo.si.edu/smbc/Research/Cacao/cacao.html>, viewed 9 December 2001.

15. Center for New Crops and Plants Products, op. cit. note 11.

16. FAO, op. cit. note 7.

17. U.N. Conference on Trade and Development, "Cocoa Talks Resume in Geneva as World Prices Soar," press release, 23 February 2001, at <www.unctad.org/en/press/pr0104en.htm>, viewed 4 January 2001; FAO, op. cit. note 7.

18. FAO, op. cit. note 7.

19. ICCO, op. cit. note 13; Gray, op. cit. note 3.

20. Gray, op. cit. note 3.

21. Jim Lobe, "Chocolate Firms Agree to Fight Cocoa Child Slavery," *Inter Press Service*, 4 October 2001.

22. International Labor Organization (ILO), "Chocolate Industry Signs Accord to End Child Labor," at <us.ilo.org/news/focus/0110/FOCUS-3.HTML>, viewed 19 December 2001; "Chocolate Industry Agreement," 1 October 2001, at <www.senate.gov/~harkin/specials/20011001-chocolate.cfm>; ILO, "C182, Worst Forms of Child Labour Convention 189," at <www.ilo.org/public/english/standards/ipec/ratification/convention/text.htm>, viewed 15 December 2001.

23. Jeffrey A. McNeely and Sara J. Scherr, "Common Ground, Common Future: How Eco-agriculture Can Help Feed the World and Save Biodiversity," <www.futureharvest.org/pdf/biodiversity_report.pdf>, viewed 15 December 2001; Rice and Greenberg, op. cit. note 5, pp. 167–73.

## SUGAR AND SWEETENER USE GROWS
(pages 32–33)

1. U.N. Food and Agriculture Organization (FAO), *FAOSTAT Statistics Database*, at <apps.fao.org>, updated 7 November 2001; estimates for 2000 and 2001 based on U.S. Department of Agriculture (USDA), *Production, Supply, and Distribution*, electronic database, Washington, DC, updated December 2001.

2. FAO, op. cit. note 1. Note that this is the uncorrected per capita value, which does not account for food waste. There are 4 calories per gram of sugar.

3. FAO, op. cit. note 1; Gladys C. Moreno Garcia, statistician, e-mail to author, 25 January 2002.

4. FAO, op. cit. note 1.

5. F. O. Licht Commodity Analysis, *World HFS Production 1987–2001*, e-mail to author, 2 November 2001.

6. Ibid.

7. USDA, op. cit. note 1.

8. F. O. Licht Commodity Analysis, op. cit. note 5.

9. High-fructose syrup consumption from ibid.; sugar consumption from USDA, op. cit. note 1.

10. Consumption in 1999 from SRI Chemical and Health Business Services, *High Intensity Sweeteners Report, Chemical Economics Handbook*, July 2000, according to Sebastian Bizzari, SRI, discussion with author, 31 October 2001; growth trend from Economic Research Service, USDA, *Per Capita Food Consumption Data System: Sugars/Sweeteners*, at <www.ers.usda.gov/Data/Food Consumption/Spreadsheets/sweets.xls>, viewed 16 December 2001.

11. Calculation based on data in SRI Chemical and Health Business Services, op. cit. note 10.

12. High-intensity sweeteners are either noncaloric due to the body's inability to process them or they are so concentrated that only a minute quantity is needed, so they contribute marginally to a person's energy intake.

13. U.S. data from Andrew Laumbach, Office of Food Additive Safety, U.S. Food and Drug Administration, Washington, DC, letter to author, 17 December 2001; Canada from Réjean Fiset, Canadian Food Inspection Agency, Ottawa, letter to author, 8 January 2002.

14. FAO, op. cit. note 1.

15. Ibid.

16. World Health Organization, *Diet, Nutrition, and the Prevention of Chronic Diseases* (Geneva: 1990), p. 113.

17. FAO, op. cit. note 1.

18. Ibid.

19. National Research Council, Committee on Diet and Health, *Diet and Health: Implications for Reducing Chronic Disease Risk* (Washington, DC: 1989), p. 638.

20. Ibid., p. 639.

21. Judy Putnam et al., "Per Capita Food Supply Trends: Progress Toward Dietary Guidelines," *FoodReview*, September–December 2000, p. 12. The USDA's recommended maximum sweetener intake is 180 calories (45 grams) per day.

22. Gary Gardner and Brian Halweil, *Underfed and Overfed: The Global Epidemic of Malnutrition*, Worldwatch Paper 150 (Washington, DC: Worldwatch Institute, March 2000).

## IRRIGATED AREA RISES (pages 34–35)

1. U.N. Food and Agriculture Organization (FAO), "Irrigation" and "Land Use" data collections, *FAOSTAT Statistics Database*, at <apps.fao.org>, updated 10 July 2001.

2. Ibid.; population data from U.S. Bureau of the Census, *International Data Base*, electronic database, Suitland, MD, updated 10 May 2000.

3. FAO, op. cit. note 1.

4. Ibid.

5. Ibid.

6. Ibid.

7. Ibid.

8. Ibid.; proportion of goods from Ruth S. Meinzen-Dick and Mark W. Rosegrant, eds., "Overview," in *Overcoming Water Scarcity and Quality Constraints* (Washington, DC: International Food Policy Research Institute (IFPRI), October 2001).

9. Peter H. Gleick, *The World's Water 2000–2001* (Washington, DC: Island Press, 2000), p. 64; Stanley Wood, Kate Sebastian, and Sara J. Scherr, "Agroecosystems," in *Pilot Analysis of Global Ecosystems* (Washington, DC: IFPRI and World Resources Institute, 2000), p. 6.

10. Meinzen-Dick and Rosegrant, op. cit. note 8.

11. Grain production from U.S. Department of Agriculture, *Production, Supply, and Distribution*, electronic database, Washington, DC, updated December 2001; FAO, op. cit. note 1.

12. Lester R. Brown, *Eco-Economy* (New York: W.W. Norton & Company, 2001), pp. 43–44; World Bank, *China: Agenda for Water Sector Strategy for North China* (Washington, DC: April 2001).

13. Sandra Postel, *Pillar of Sand* (New York: W.W. Norton & Company, 1999), pp. 73–74.

14. Ibid., p. 77.

15. Ibid.

16. Ibid., p. 92.

17. Wood, Sebastian, and Scherr, op. cit. note 9, pp. 57–58; Postel, op. cit. note 13, p. 19.

18. Sandra Postel, "Redesigning Irrigated Agriculture," in Worldwatch Institute, *State of the World 2000* (New York: W.W. Norton & Company, 2000), p. 51.

19. Sandra Postel et al., "Drip Irrigation for Small Farmers: A New Initiative to Alleviate Hunger and Poverty," *Water International*, March 2001, p. 5.

20. Ibid.

21. Ibid., p. 12.

22. Ibid., p. 8.

## FOSSIL FUEL USE INCHES UP (pages 38–39)

1. Based on United Nations, *World Energy Supplies, 1950–74* (New York: 1976), on BP, *BP Statistical Review of World Energy June 2001* (London: Group Media & Publishing, 2001), on International Energy Agency (IEA), "Oil Market Report" (Paris: 18 January 2002), on idem, "Monthly Natural Gas Survey" (Paris: October 2001), on U.S. Department of Energy (DOE), Energy Information Administration, *International Energy Outlook* (Washington, DC: March 2001), on idem, *Monthly Energy Review* (Washington, DC: January 2002), on "Coal Price Rise Acceptable," *China Daily*, 3 January 2002, and on "China Exports Record Amount of Coal This Year," *Xinhua*, 24 December 2001.

2. Based on United Nations, op. cit. note 1, on BP, op. cit. note 1, on IEA, "Oil Market Report," op. cit. note 1, on idem, "Monthly Natural Gas Survey," op. cit. note 1, on DOE, *International Energy Outlook*, op. cit. note 1, on idem, *Monthly Energy Review*, op. cit. note 1, on *China Daily*, op. cit. note 1, and on *Xinhua*, op. cit. note 1.

3. Based on United Nations op. cit. note 1, on BP, op. cit. note 1, on IEA, "Oil Market Report," op. cit. note 1, on idem, "Monthly Natural Gas Survey," op. cit. note 1, on DOE, *International Energy Outlook*, op. cit. note 1, on idem, *Monthly Energy Review*, op. cit. note 1, on *China Daily*, op. cit. note 1, and on *Xinhua*, op. cit. note 1.

4. IEA, "Oil Market Report," op. cit. note 1.

5. Ibid.

6. Ibid.
7. Ibid.
8. IEA, "Monthly Natural Gas Survey," op. cit. note 1.
9. Ibid.
10. Ibid.
11. DOE, *International Energy Outlook*, op. cit. note 1.
12. DOE, *Monthly Energy Review*, op. cit. note 1.
13. *China Daily*, op. cit. note 1; *Xinhua*, op. cit. note 1.
14. Jonathan E. Sinton, "Accuracy and Reliability of China's Energy Statistics," *China Economic Rreview*, vol. 12 (2001), pp. 373–83.
15. IEA, *World Energy Outlook: 2001 Insights* (Paris: Organisation of Economic Co-operation and Development, 2001).
16. Ibid.
17. Ibid.
18. Ibid.
19. Ibid.
20. Ibid.

## NUCLEAR POWER UP SLIGHTLY
(pages 40–41)

1. Installed nuclear capacity is defined as reactors connected to the grid as of 31 December 2001, and is based on Worldwatch Institute database compiled from statistics from the International Atomic Energy Agency and press reports primarily from *Associated Press, Reuters, Agence France-Presse, World Nuclear Association (WNA) News Briefing*, and Web sites.
2. Worldwatch Institute database, op. cit. note 1.
3. "Russia's Newest Nuclear Power Plant Turned Up to Full Capacity," *Associated Press*, 6 September 2001.
4. Worldwatch Institute database, op. cit. note 1.
5. Ibid.
6. Ibid.
7. "U.S. Lawmakers See Resurgence in Nuclear Power," *Reuters*, 27 March 2001.
8. "Global Atomic Agency Confesses Little Can Be Done to Safeguard Nuclear Plants," *Associated Press*, 17 September 2001.
9. "Belgium Agrees to Give Up Nuclear Power: Official," *Agence France-Presse*, 9 October 2001; "Germany: An Agreement on Phasing Out the Production of Nuclear Power," *WNA News Briefing*, 6–12 June 2001.
10. "New U-Turn on Nuclear Powerplant," (London) *Mail on Sunday*, 11 January 2002.
11. "Sweden," *WNA News Briefing*, 5–11 December 2001.
12. "Russia to Build Ten Nuclear Reactors in Ten Years—Ryabev," *Itar-Tass News Agency*, 14 November 2001.
13. "Ukraine Turns to Russia for Help in Completing Nuclear Reactors," *Associated Press*, 29 November 2001.
14. Chihiro Kamisawa, "Referendum at Kariwa Village: A Strong 'No' to MOX Program," *Nuke Info Tokyo*, July/August 2001; "Mie Town Votes Against Urging Firm Build Nuclear Plant," *Kyodo News*, 18 November 2001.
15. "Oma Nuclear Plant Plan 'Temporarily Suspended'," *Nuke Info Tokyo*, January/February 2002.
16. Worldwatch Institute database, op. cit. note 1
17. "China Begins Work on 6th Nuclear Power Plant," *Kyodo News*, 7 January 2002.
18. Worldwatch Institute database, op. cit. note 1.
19. "Taiwan Decides Against Nuclear Power Plant Referendum," *Kyodo News*, 10 August 2001.
20. "Taiwan: The Future of the Lungmen Nuclear Power Plant Project," *WNA News Briefing*, 5–11 December 2001.
21. "Russia: Seven Men Were Arrested Attempting to Sell," *WNA News Briefing*, 5–11 December 2001; "Turkey: Police Arrested Two Men," *WNA News Briefing*, 7–13 November 2001.

## WIND ENERGY SURGES (pages 42–43)

1. Worldwatch Institute preliminary estimate based on figures from European Wind Energy Association (EWEA), "Another Record Year for European Wind Power," press release (London: 18 February 2002), on Birger Madsen, BTM Consult, e-mail to author, 14 February 2002, on "Windicator," *Windpower Monthly*, January 2002, and on American Wind Energy Association (AWEA), "U.S. Wind Industry Installs Nearly 1,700 Megawatts in 2001," *Wind Energy Weekly*, 18 January 2002. Historical data from BTM Consult, *International Wind Energy Development: World Market Update, 1999* (Ringkobing, Denmark: March 2000.)
2. Worldwatch preliminary estimate based on sources in note 1.
3. Ibid.
4. Worldwatch estimate based on EWEA, op. cit. note 1, and on Madsen, op. cit. note 1.
5. EWEA, op. cit. note 1.
6. Ibid.

# Notes

7. Ibid.
8. "Company Profile: Gamesa Eolica and Energia," *Wind Directions*, January 2002, p. 12.
9. EWEA, op. cit. note 1.
10. Jorn Ruby, "New Trading System to Generate More Competition?" *New Energy*, June 2000, p. 22.
11. Preben Maegaard, Director, Folkecenter for Renewable Energy, "Disaster for Danish Renewable Energy: New Government Cancels All Renewable Energy Programmes," unpublished, February 2002.
12. EWEA, op. cit. note 1.
13. AWEA, op. cit. note 1.
14. Mike O'Bryant, "United States Winds Up Record Year," *Windpower Monthly*, January 2002, p. 21.
15. AWEA, "Wind Energy Tax Credit Extended for Two Years," press release (Washington, DC: 8 March 2002).
16. Karl Royce, "Brazil Regulator Launches Bonanza," *Windpower Monthly*, January 2002, p. 18.
17. Ibid.
18. Figure of $7 billion is author's estimate based on capacity installed.
19. "GE Power Systems to Purchase Enron Wind Corp. Assets," *Wind Energy Weekly*, 22 February 2002.

## SOLAR CELL USE RISES QUICKLY
(pages 44–45)

1. Paul Maycock, *PV News*, February 2002, and Paul Maycock, PV Energy Systems, Warrenton, VA, letter to author, 28 February 2002.
2. Paul Maycock, *PV News*, various issues.
3. Cumulative PVs installed from ibid., with the conservative assumption of a 10-percent decline in installed PV capacity each year.
4. Peter Holihan, *Technology, Manufacturing, and Market Trends in the U.S. and International Photovoltaics Industry* (Washington, DC: U.S. Department of Energy, Energy Information Administration, no date) at <www.eia.doe.gov/cneaf/solar.renewables/rea_issues/solar.html>, viewed 4 March 2002.
5. Bob Johnson, Strategies Unlimited, Mountain View, CA, discussion with author, 29 February 2002.
6. Maycock, op. cit. note 2.
7. Paul Maycock, "Boomer," *PV News*, January 2002.
8. Maycock, letter to author, op. cit. note 1.
9. Maycock, op. cit. note 7.
10. Maycock, letter to author, op. cit. note 1.
11. Ibid.
12. Ibid.
13. "Solar Finally Gets Practical," *Los Angeles Times*, 9 February 2002; Jim Carlton, "Ballot Measures Would Give Solar Industry Big Boost," *Wall Street Journal*, 6 November 2001; "San Francisco Votes to Produce Its Own Solar and Wind Power," *New York Times*, 9 November 2001.
14. Maycock, op. cit. note 7.
15. Maycock, op. cit. note 2.
16. Paul Maycock, PV Energy Systems, Warrenton, VA, discussion with author, 28 February 2002.
17. Ibid.; description of thin films from Peter Fairley, "Solar on the Cheap," *Technology Review*, January/February 2002, pp. 48–53.
18. Maycock, op. cit. note 16.
19. David Berry, "Sunshine in the City: Photovoltaics for Local Power," *PM: Public Management*, November 2001, pp. 15–19.
20. Stewart Boyle, "Daze in the Sun," *Tomorrow Magazine*, June 2001, pp. 64–67.
21. Estimate of 1.7 billion for population without electrical access in 2000 from Anil Cabraal and Kevin Fitzgerald, "PV for Rural Electrification within Restructured Power Sectors in Developing Countries," Asia Alternative Energy Program, World Bank, unpublished manuscript, updated 5 March 2002. See also David Lipschultz, "Solar Power is Reaching Where Wires Can't," *New York Times*, 9 September 2001.
22. Dollar estimate from Anil Cabraal, Senior Energy Specialist, Asia Alternative Energy Program, World Bank, e-mail to author, 6 March 2002; number of systems from Anil Cabraal, "Building on Experience: Assuring Quality in the World Bank/GEF-Assisted China Renewable Energy Development Project," paper presented at the 16th European Photovoltaics Conference, Glasgow, Scotland, May 2000.
23. Michael T. Eckhart, Jack L. Stone, and Keith Rutledge, "Financing PV Growth," *PV Engineering Handbook* (London: Wiley & Sons, in press).
24. Neville Williams, President, Solar Electric Light Company, Chevy Chase, MD, discussion with author, 6 March 2002.

## COMPACT FLUORESCENTS SET RECORD
(pages 46–47)

1. Nils Borg, International Association for Energy Efficient Lighting (IAEEL), e-mail to author, 11 January 2002.
2. Nils Borg, IAEEL, e-mail to author, 26 January 2000.
3. Data for 1988–89 from Evan Mills, Lawrence Berkeley Laboratory, Berkeley, CA, letter to Worldwatch, 3 February 1993; 1990–2001 estimate from Borg, op. cit. note 1.
4. Worldwatch estimate based on a 15-percent decay in existing compact fluorescent lamp (CFL) stock per year, and equivalent lumen output derived from 15-watt CFLs and 60-watt incandescents.
5. Worldwatch estimate based on 15-watt CFLs replacing 60-watt incandescents, used for 4 hours per day, and where an "average-sized coal fired power plant" is a 440-megawatt electric output, operating 80 percent of the time (80 percent availability).
6. Worldwatch estimate based on 15-percent decay in North American CFL stock per year, 15-watt CFLs replacing 60-watt incandescents, 4 hours of lighting per day, and 0.1839 short tons of carbon and 0.0036 short tons of sulfur dioxide per 1,000 kilowatt-hours of electricity generated in the United States. While the estimate expresses the savings from installing CFLs instead of incandescents, it is difficult to determine how many CFLs are literally used in place of incandescents. Electricity consumption and emissions data from U.S. Department of Energy (DOE), Energy Information Administration (EIA), *Electric Power Annual 1999 Volume II* (Washington, DC: 1999).
7. Worldwatch estimate based on a 10,000-hour 15-watt CFL costing 300 baht replacing a 750-hour 60-watt incandescent costing 20 baht, with lamps used 4 hours per day and electricity costing 3 baht per kilowatt-hour; Peter du Pont, Chief Technical Advisor, Danish Energy Management A/S, Bangkok, Thailand, e-mail to author, 20 January 2002.
8. Worldwatch estimate based on a 5-percent discount rate in the net present value calculation; energy and bulb prices from du Pont, op. cit. note 7.
9. Priority 4, Clear Energy and Transportation, activity 4-9 China Green Lights Programme, in "The Priority Programme for China's Agenda 21 (Revised and Expanded Version)," at <www.zhb.gov.cn/english/SD/21cn/priority/cn4-9.htm>.
10. Mark Levine, "The Green Lights Program of China," *World Energy Efficiency News*, September 1997.
11. Borg, op. cit. note 1.
12. Russell Sturm, International Finance Corporation, e-mail to author, 7 January 2002.
13. Ibid.
14. Ibid.
15. Ibid.
16. Barry Bredenkamp, Bonesa, e-mail to author, 9 January 2002.
17. Sturm, op. cit. note 12.
18. Nicholas Iacobucci, CFL Product Manager, General Electric Lighting North America, e-mail to author, 29 January 2002; Tracey Rembert, "Green Living Eco-Home," *E Magazine*, July-August 1996.
19. Estimated U.S. electric power sector mercury emissions for 1997 were 52 tons, equating to 0.01495 milligrams of mercury per kilowatt-hour across all U.S. generators in 1997. Over the average 10,000-hour life span, a 15-watt CFL (replacing a 60-watt incandescent) will save 6.73 milligrams of mercury emissions from the electric power sector. "Analysis of Strategies for Reducing Multiple Emissions from Electric Power Plants: Sulfur Dioxide, Nitrogen Oxides, Carbon Dioxide, and Mercury and a Renewable Portfolio Standard"(Washington, DC: EIA, DOE, July 2001); DOE, *Annual Energy Outlook 2000* (Washington, DC: December 1999).

## GLOBAL TEMPERATURE CLOSE TO A RECORD (pages 50–51)

1. J. Hansen et al., "Global Land-Ocean Temperature Index in .01 C," at <www.giss.nasa.gov/data/update/gistemp>, viewed 25 January 2002.
2. Ibid.
3. World Meteorological Organization, "WMO Statement on the Status of the Global Climate in 2001," 18 December 2001.
4. Ibid.
5. Ibid.
6. Ibid.
7. Ibid.
8. Ibid.
9. Ibid.
10. Ibid.

# Notes

11. Ibid.
12. Ibid.
13. Ibid.
14. Ibid.
15. Ibid.
16. Ibid.
17. Ibid.
18. Ibid.
19. Ibid.
20. Ibid.
21. Ibid.
22. Ibid.
23. Ibid.
24. Ibid.
25. Ibid.
26. Ibid.
27. Ibid.
28. Ibid.
29. Ibid.
30. Ibid.
31. J. T. Houghton et al., eds., *Climate Change 2001: The Scientific Basis*, Contribution of Working Group I to the Third Assessment Report of the Intergovernmental Panel on Climate Change (Cambridge, U.K.: Cambridge University Press, 2001).
32. Ibid., p. 10.

## CARBON EMISSIONS REACH NEW HIGH
(pages 52–53)

1. Historical trends and preliminary 2001 estimate based on G. Marland, T.A. Boden, and R.J. Andres, Carbon Dioxide Information Analysis Center, Oak Ridge National Laboratory, "Global, Regional, and National Annual $CO_2$ Emissions from Fossil-Fuel Burning, Cement Production, and Gas Flaring: 1751–1998 (revised July 2001)," at <www.cdiac.esd.ornl.gov/ndps/nps030.html>, viewed 12 January 2002, on BP, *BP Statistical Review of World Energy June 2001* (London: Group Media & Publishing, 2001), on International Energy Agency (IEA), "Oil Market Report" (Paris: 18 January 2002), on idem, "Monthly Natural Gas Survey" (Paris: October 2001), on U.S. Department of Energy (DOE), Energy Information Administration, *International Energy Outlook* (Washington, DC: March 2001), on idem, *Monthly Energy Review* (Washington, DC: January 2002), on "Coal Price Rise Acceptable," *China Daily*, 3 January 2002, and on "China Exports Record Amount of Coal This Year," *Xinhua*, 24 December 2001.
2. Based on Marland, Boden, and Andres, op. cit. note 1, on BP, op. cit. note 1, on IEA, "Oil Market Report," op. cit. note 1, on idem, "Monthly Natural Gas Survey," op. cit. note 1, on DOE, *International Energy Outlook*, op. cit. note 1, on idem, *Monthly Energy Review*, op. cit. note 1, on *China Daily*, op. cit. note 1, and on *Xinhua*, op. cit. note 1.
3. Worldwatch estimate based on Marland, Boden, and Andres, op. cit. note 1, and on BP, op. cit. note 1.
4. Ibid.
5. Ibid.
6. Ibid.
7. UN Framework Convention on Climate Change (UN FCCC), "Kyoto Protocol to the UN Framework Convention on Climate Change," 11 December 2001, at <www.unfcc.int>.
8. Worldwatch estimate based on Marland, Boden, and Andres, op. cit. note 1, and on BP, op. cit. note 1.
9. Marland, Boden, and Andres, op. cit. note 1.
10. Based on ibid., on BP, op. cit. note 1, on IEA, "Oil Market Report," op. cit. note 1, on idem, "Monthly Natural Gas Survey," op. cit. note 1, on DOE, *International Energy Outlook*, op. cit. note 1, on idem, *Monthly Energy Review*, op. cit. note 1, on *China Daily*, op. cit. note 1, and on *Xinhua*, op. cit. note 1.
11. Based on Marland, Boden, and Andres, op. cit. note 1, on BP, op. cit. note 1, on IEA, "Oil Market Report," op. cit. note 1, on idem, "Monthly Natural Gas Survey," op. cit. note 1, on DOE, *International Energy Outlook*, op. cit. note 1, on idem, *Monthly Energy Review*, op. cit. note 1, on *China Daily*, op. cit. note 1, and on *Xinhua*, op. cit. note 1.
12. J. T. Houghton et al., eds., *Climate Change 2001: The Scientific Basis* (Cambridge, U.K.: Cambridge University Press, 2001).
13. C. D. Keeling and T. Whorf, Scripps Institution of Oceanography, La Jolla, CA, e-mail to author, 29 January 2002.
14. Ibid.
15. Ibid.
16. UN FCCC, "Governments Ready to Ratify Kyoto Protocol," press release (New York: 10 November 2001).
17. Ibid.

18. UN FCCC, "Kyoto Protocol: Status of Ratification," at <www.unfccc.int>, updated 6 March 2002.

19. Eric Pianin, "Bush Unveils Global Warming Plan," *Washington Post*, 15 February 2002.

20. Michael Grubb, Jean-Charles Hourcade, and Sebastian Oberthur, *Keeping Kyoto: A Study of Approaches to Maintaining the Kyoto Protocol on Climate Change* (London: Climate Strategies, July 2001).

21. "EU Agrees to Ratify Kyoto Pact by June," *International Herald Tribune Online*, 5 March 2002.

22. Geoff Winestock, "EU Ministers Support Adoption of Environmental Accord," *Wall Street Journal-Europe*, 5 March 2002; World Wildlife Fund, "Go For Kyoto," at <www.panda.org/goforkyoto/table_rat.cfm>, viewed 4 March 2002.

## CFC USE DECLINING (pages 54–55)

1. U.N. Environment Programme (UNEP), "Data on Production of CFCs (Annex A, Group I) ODP Tonnes," e-mail from Gerald Mutisya, UNEP Ozone Secretariat to author, 8 March 2002; idem, e-mail to author, 14 January 2002. The UNEP Ozone Secretariat collects data on CFC production from countries that are party to the Montreal Protocol. These numbers reflect the volume in tons of the major CFCs (CFC-11, CFC-12, CFC-113, CFC-114, and CFC-115), multiplied by their respective ozone-depleting potentials (ODPs). The ODP value is the ratio of a given compound's ability to deplete ozone compared with the ability of a similar mass of CFC-11. The UNEP data series differs from that used in previous editions of *Vital Signs*, as those were based on industry estimates discontinued after CFC production was phased out in industrial nations.

2. UNEP, "Data on Production," op. cit. note 1.

3. Ibid. Note that production in the United States and some nations in Western Europe was reported as negative between 1996 and 1999, but the negative U.S. producton is not reflected in the figure. (A nation's total CFC production can be counted as negative when CFCs—either stockpiles of never-used material or CFCs recovered from appliances—are destroyed. Negative production also occurs when stockpiled CFCs are used as feedstock to create other chemicals, although this is less common.)

4. "CFCs Continued to be Allowed for Certain Uses," *Chemical Market Reporter*, 19 November 2001, p. 9; Ashley Woodcock, "CFCs for Inhalers—The Beginning of the End?" *The Lancet*, 23–30 December 2000, p. 2166.

5. Michael Graber, UNEP Ozone Secretariat, e-mail to author, 19 January 2001.

6. Ibid.

7. Ibid.

8. Bill Schmitt, "Next Generation Debuts, as HCFCs Drift Away," *Chemical Week*, 8 August 2001, pp. 27–29.

9. D'Vora Ben Shaul, "'Greenfreeze' Promises Cold Comfort For All—In a Very Literal Way," *Jerusalem Post*, 29 November 2001.

10. Ibid.

11. Eric Lai, "CFCs? What CFCs? Few Cars Affected by Law Banning Them," *Toronto Star*, 4 August 2001.

12. "CFC Replacements Near Halfway Mark," *Consulting-Specifying Engineer*, 1 May 2001, p. 17.

13. Environmental Investigation Agency, *Unfinished Business: The Continued Illegal Trade in Ozone Depleting Substances and the Threat Posed to the Montreal Protocol Phase-out* (London: October 2001), at <www.eia-international.org>, viewed 10 January 2002.

14. UNEP, "Illegal Trade in Ozone Depleting Substances: Is There A Hole in the Montreal Protocol?" *OzonAction* Special Supplement, October 2001.

15. Linda Baker, "The Hole in the Sky," *E Magazine*, November/December 2000, pp. 34–40.

16. Fred Pearce, "Ozone Unfriendly: A Quartet of 'Green' Chemicals Now Face a Total Ban," *New Scientist*, 20 October 2001, p. 17.

17. Molina cited in ibid.

18. "Ozone Hole Report," *Chemical Week*, 31 October 2001.

19. "NOAA and NASA Keeping Close Watch on Ozone Hole Over Antarctica," *Bulletin of the American Meteorological Society*, December 2001, pp. 2883–85.

20. Hemel Hempstead, "Scientists Predict Ozone Hole Will Close Up in 50 Years," *Appropriate Technology*, January–March 2001, p. 23; Mark Henderson, "Ozone Hole Will Heal in 50 Years, Say Scientists," *The Times*, 4 December 2000.

21. Jimmy Langman, "Under the Hole in the Sky," *Newsweek* (International Edition), 3 December 2001, p. 62.

# Notes

## ECONOMIC GROWTH FALTERS
(pages 58–59)

1. Worldwatch estimate, based on Angus Maddison, *The World Economy: A Millennial Perspective* (Paris: Organisation for Economic Co-operation and Development, 2001), pp. 272–321, with updates from International Monetary Fund (IMF), *World Economic Outlook Database* (Washington, DC: December 2001) with IMF figures for China multiplied by 0.759, following Maddison, e-mail to author, 16 January 2002. All gross domestic product estimates are converted from other currencies to dollars on the basis of purchasing power parities.
2. Ibid.
3. IMF, *World Economic Outlook* (Washington, DC: December 2001), p. 3.
4. IMF, op. cit. note 1.
5. Ibid.
6. Ibid.
7. Worldwatch, op. cit. note 1.
8. IMF, op. cit. note 1.
9. Worldwatch, op. cit. note 1.
10. Ibid.
11. Causality from IMF, op. cit. note 3, p. 3; prices from Department of Energy, Energy Information Administration, "U.S. Petroleum Prices," at <www.eia.doe.gov/oil_gas/petroleum/info_glance/prices.html>, viewed 14 January 2002.
12. IMF, op. cit. note 3, p. 3.
13. Worldwatch estimates, based on the Wilshire 5000 index, which is calibrated so that one point equals $1 billion in U.S. stock market value. Data from Wilshire Associates, at <www.wilshire.com/Indexes/calculator>, viewed 14 January 2002.
14. Microsoft Corporation, MSN MoneyCentral, "International Indexes," at <moneycentral.msn.com/investor/market/foreign.asp>, viewed 14 January 2002.
15. IMF, op. cit. note 3, p. 18.
16. Business Cycle Dating Committee, National Bureau of Economic Research, "The Business-Cycle Peak of March 2001," Cambridge, MA, 26 November 2001.
17. Deaths from Dennis Normile, "Quake Builds Strong Case for Codes," *Science*, 27 January 1995, p. 444; effects from IMF, op. cit. note 3, p. 16.
18. "US Recession Raises Global Fears," *BBC News*, 18 November 2001.
19. Worldwatch, op. cit. note 1. Figure is for Western Europe, Australia, Canada, New Zealand, and the United States.
20. Worldwatch, op. cit. note 1.
21. Maddison, *The World Economy*, op. cit. note 1, p. 298; poverty figures are from UNICEF, cited in World Bank, *East Asia Regional Overview* (Washington, DC: September 1999), and show the change from mid-1997 to mid-1998.
22. U.S. Bureau of Labor Statistics, data series LFS11000000 and LFS40000000, at <data.bls.gov/cgi-bin/surveymost?lf>, 14 January 2002.

## TRADE SLOWS (pages 60–61)

1. International Monetary Fund (IMF), *World Economic Outlook Database* (Washington, DC: December 2001); Figure 1 is based on ibid. for 1970–2001, and on IMF, *International Financial Statistics*, electronic database, Washington, DC, November 2001 for 1950–69. For areas of overlap (goods exports for 1970–2001), these two sources differ, but never by more than 2.5 percent.
2. IMF, *World Economic Outlook Database*, op. cit. note 1.
3. IMF, *World Economic Outlook* (Washington, DC: December 2001), p. 3.
4. IMF, *International Financial Statistics*, op. cit. note 1.
5. IMF, *World Economic Outlook Database*, op. cit. note 1.
6. Ibid.
7. IMF, *Balance of Payment Statistics Yearbook*, Part 2 (Washington, DC: 2000), pp. 24–34.
8. IMF, *World Economic Outlook Database*, op. cit. note 1. All gross domestic product estimates are converted from other currencies to dollars on the basis of purchasing power parities.
9. Based on the Export Unit Value Index, from IMF, *International Financial Statistics*, op. cit. note 1.
10. Ibid.
11. Charlotte Denny, "Fresh Fudge Recipe Is a Fusion of French and Indian Cookery," (London) *Guardian*, 15 November 2001.
12. World Trade Organization, *Trading into the Future*, 2nd ed. (Geneva: 1999), p. 51.
13. John Audley, *New Rules in International Trade* (Washington, DC: Carnegie Endowment for International Peace, 2001).

14. Ibid.
15. Ibid.
16. Ibid.

## FOREIGN DEBT FALLS IN DOLLAR TERMS
(pages 62–63)

1. World Bank, *Global Development Finance 2001*, advance release, electronic database, Washington, DC, 2001, with updates from idem, *Global Development Finance 2001*, vol. 1 (Washington, DC: 2001), pp. 246–47.
2. Ibid.
3. World Bank, *Global Development Finance 2001*, vol. 1, op. cit. note 1, p. 246.
4. International Monetary Fund (IMF), *International Financial Statistics* (Washington, DC: November 2001), pp. 328–29.
5. Worldwatch estimate, based on ibid.
6. Mahn-Je Kim, "The Republic of Korea's Successful Economic Development and the World Bank," in Devesh Kapur, John P. Lewis, and Richard Webb, eds., *The World Bank: Its First Half Century*, vol. 2 (Washington, DC: Brookings Institution Press, 1997), p. 25.
7. Karin Lissakers, *Banks, Borrowers, and the Establishment: A Revisionist Account of the International Debt Crisis* (New York: BasicBooks, 1991), pp. 60–83.
8. World Bank, *Global Development Finance*, vol. 1, op. cit. note 1, pp. 272–73.
9. Ibid., pp. 246, 272.
10. Based on Christian Suter, *Debt Cycles in the World-Economy: Foreign Loans, Financial Crises, and Debt Settlements, 1820–1990* (Boulder, CO: Westview Press, 1992), pp. 70–71, and on Worldwatch estimates, which are based on World Bank, *Global Development Finance*, vol. 1, op. cit. note 1, pp. 157–82, on idem, *Annual Report* (Washington, DC: various years), notes to IBRD and IDA financial statements, and on IMF, *Annual Report* (Washington, DC: various years), notes to financial statements. Figure 2 includes defaults, consolidation periods (agreed periods during which no payments are made), and instances of countries going into "nonaccrual" status with the World Bank or "overdue" status with the IMF. Five defaults—Bulgaria's in 1932, China's in 1939, Czechoslovakia's in 1960, East Germany's in 1949, and Russia's in 1918—are treated as having terminated at the end of 1989, when the organization representing holders of these nations' bonds, the London-based Corporation of Foreign Bondholders, dissolved.
11. Joan M. Nelson, "The Politics of Pro-Poor Adjustment," in Joan M. Nelson, ed., *Fragile Coalitions: The Politics of Economic Adjustment* (Oxford: Transaction Books, 1989), p. 111.
12. Wilfredo Cruz and Robert Repetto, *The Environmental Effects of Stabilization and Structural Adjustment Programs: The Philippines Case* (Washington, DC: World Resources Institute, 1992), p. 50.
13. World Bank, *Global Development Finance*, vol. 1, op. cit. note 1, pp. 270–71.
14. For evidence of "defensive lending," see Nancy Birdsall, Stijn Claessens, and Ishac Diwan, "Will HIPC Matter? The Debt Game and Donor Behavior in Africa," WIDER Development Conference on Debt Relief, Helsinki, 17 August 2001, at <www.wider.unu.edu/conference/conference-2001-2/plenary%20papers/Birdsall%20et%20al.pdf>, and see Dilip Ratha, *Demand for World Bank Lending*, Policy Research Working Paper 2652 (Washington, DC: World Bank, 2001).
15. William Easterly, *The Elusive Quest for Growth: Economists' Adventures and Misadventures in the Tropics* (Cambridge, MA: The MIT Press, 2001), pp. 125–27.
16. The World Bank criteria are gross national product per capita of less than $885 per year and present value of debt equal to 18 months or more of export earnings. Count of 47 includes: all 42 HIPC-eligible nations except Bolivia (whose GNP per capita now exceeds $885), listed at World Bank, *Debt Initiative for Heavily Indebted Poor Countries*, at <www.worldbank.org/hipc>, viewed 23 December 2001; Cambodia, the Kyrgyz Republic, Nigeria, and Pakistan, which also meet the criteria; and Afghanistan and Indonesia, which are listed as "severely indebted low income" in idem, *Global Development Finance*, vol. 1, op. cit. note 1, pp. 141–42.
17. Figure of 42 is from World Bank, op. cit. note 16; figure of 55 percent includes debt cancellation offers made by Group of Seven and other governments over and above HIPC, and is a Worldwatch estimate, based on David Andrews et al., *Debt Relief for Low-Income Countries: The Enhanced HIPC Initiative*, Pamphlet Series No. 51 (Washington, DC: IMF, 2000), on *Global Development Finance*, vol. 1, op. cit. note 1, pp. 142–44, and on Horst Köhler and James D. Wolfensohn, "Debt

# Notes

Relief for the Poorest Countries: Milestone Achieved," joint statement, background charts, World Bank and IMF, Washington, DC, 22 December 2000. Estimate assumes all eligible countries participate.

18. Worldwatch estimate based on 55-percent figure and on Daniel Cohen, *The HIPC Initiative: True and False Promises*, Working Paper (Paris: Organisation for Economic Co-operation and Development, Development Centre, 2000).

## METALS EXPLORATION DROPS SHARPLY
(pages 64–65)

1. Investment number based on Metals Economics Group (MEG), *Strategic Report*, November/December 1991; idem, *Strategic Report*, September/October 1992; idem, *Strategic Report*, September/October 1993; idem, "Major Increase in Junior Spending," press release (Halifax, NS, Canada: 14 October 1994); idem, "1996 Exploration Surges Upward," press release (Halifax, NS, Canada: 16 October 1996); idem, "Latin America Tops Exploration Spending for the Fourth Year," press release (Halifax, NS: 16 October 1997); idem, "A 31% Decrease in 1998 Exploration Budgets," press release (Halifax, NS, Canada: 20 October 1998), idem, "A 23% Decrease in 1999 Exploration Budgets," press release (Halifax, NS, Canada: 20 October 1999), and idem, "Exploration Spending Drops to its Lowest Level in Nine Years," press release (Halifax, NS, Canada: 1 November 2001). The calculations of this Canadian consultancy group are based on budgets reported by major mining companies that represent 80–90 percent of worldwide exploration expenditures for precious, base, and other nonferrous hard metals. In 2001, this totaled $2 billion; junior companies accounted for another $20 million.

2. MEG, "A 31% Decrease in 1998 Exploration Budgets," op. cit. note 1.

3. MEG, "Exploration Spending Drops," op. cit. note 1.

4. John Culjak, MEG, Halifax, NS, Canada, discussion with author, 9 January 1998.

5. John Culjak, MEG, Halifax, NS, Canada, discussion with author, 15 January 2002; gold price from Kitco Precious Metals, at <www.kitco.com/charts/historicalgold.html>, viewed 10 January 2002, converted to 2000 dollars using U.S Commerce Department deflator series.

6. Culjak, op. cit. note 5.

7. Based on MEG, *Strategic Report*, 1991, op. cit. note 1; idem, *Strategic Report*, 1992, op. cit. note 1; idem, "Latin America Tops," op. cit. note 1.

8. MEG, "Exploration Spending Drops," op. cit. note 1.

9. Latin America and Figure 2 from MEG, "Exploration Spending Drops," op. cit. note 1.

10. MEG, *Strategic Report*, September/October 2000 (Halifax, NS, Canada: 2000), p. 5.

11. MEG, "Exploration Spending Drops," op. cit. note 1.

12. Ibid.

13. Ibid.

14. Ibid.

15. Douglas Jehl, "Gold Miners Eager for Bush to Roll Back Clinton Rules," *New York Times*, 16 August 2001; Michael McCabe, "Easing of Rules a Boon for Hard-Rock Mining; Clinton Environmental Regulations Being Erased," *San Francisco Chronicle*, 28 December 2001.

16. Matthew Green, "Mining Giant Treads Fine Line in Madagascar Forest," *Reuters*, 19 December 2001; "Mining Companies Invade Peru's Andean Cloud Forests," *Environment News Service*, 17 August 2001; Simon Denyer, "Mining Drives Congo's Gorillas Close to Extinction," *Reuters*, 10 May 2001.

17. Michael Ross, *Extractive Sectors and the Poor* (Boston, MA: Oxfam America, October 2001); Jeffrey D. Sachs and Andrew M. Warner, "Natural Resource Abundance and Economic Growth," Development Discussion Paper No. 517a (Cambridge, MA: Harvard Institute for International Development, 1995).

18. Roger Moody, "The Lure of Gold—How Golden Is the Future?" *Panos Media Briefing No. 19* (London: Panos Institute, May 1996); Saleem Ali and Larissa Behrendt, "Mining and Indigenous Rights," *Cultural Survival Quarterly*, spring 2001, pp. 6–8.

19. "Local Residents Express Environmental Concern Over Gold Mine," *Reuters*, 3 January 2002; Jason Vest, "Rivers of Cyanide," *In These Times*, 17 April 2000, pp. 21–22.

20. International Labor Organization, "Sectoral Activities: Mining," information sheet, at <www.ilo.org/public/english/dialogue/sector/sectors/mining.htm#Heading2>, viewed 14 January 2002.

21. Earle Amey, Commodity Specialist, U.S. Geological Survey, discussion with author, 10 February 2000; "Gold: Declining Value Sends Shock Waves Through Africa," *UN Wire*, 21 June 1999.

## METALS PRODUCTION CLIMBS
(pages 66–67)

1. Worldwatch estimate, based on Grecia Matos, Mineral and Material Specialist, U.S. Geological Survey (USGS), Reston, VA, e-mail to author, 20 September 2001, on USGS, *Minerals Yearbook* (Reston, VA: various years), on idem, *Mineral Commodity Summaries* (Reston, VA: various years), and on United Nations, *Industrial Commodity Statistics Yearbook* (New York: various years). All data are for primary metals production only, except for data for aluminum and magnesium, which include some secondary production.
2. Ibid.
3. Based on a 218-ton truck that is 13 meters in length; Earth's circumference at the equator is estimated at 41,000 kilometers.
4. Matos, op. cit. note 1; USGS, "Iron and Steel," *Mineral Commodity Summaries*, January 2002.
5. Worldwatch estimate, op. cit. note 1; U.S. Bureau of the Census, *International Data Base*, electronic database, Suitland, MD, updated 10 May 2000.
6. Worldwatch estimate, op. cit. note 1; Census Bureau, op. cit. note 5.
7. Worldwatch estimate, op. cit. note 1; gross world product data from Angus Maddison, *The World Economy: A Millennial Perspective* (Paris: Organisation for Economic Co-operation and Development, 2001), pp. 272–321, with updates from International Monetary Fund, *World Economic Outlook Database* (Washington, DC: December 2001).
8. Copper from USGS, "Copper," *Mineral Commodity Summaries*, January 2002; lead from USGS, *Minerals Yearbook—2000* (Reston, VA: 2000). Data for copper are for smelter production; for lead, data are for primary refinery production.
9. USGS, *Minerals Yearbook—2000*, op. cit. note 8.
10. Steel production from USGS, op. cit. note 4; consumption from "Chapter 2. Producing and Selling Minerals," in Mining, Minerals and Sustainable Development Project (MMSD), *Final Report* (draft) (London: 4 March 2002), p. 19; aluminum from USGS, *Minerals Yearbook—2000*, op. cit. note 8.
11. MMSD, op. cit. note 10, p. 7.
12. Data from CRU International cited in "Chapter 5. Case Studies on Minerals," in MMSD, op. cit. note 10, p. 8.

13. Ibid.
14. Ibid.
15. Ibid., p. 7.
16. Ibid.
17. Kenneth Geiser, *Materials Matter* (Cambridge, MA: The MIT Press, 2001), p. 220.
18. Ibid.
19. Copper from USGS, "Recycling—Metals," *Minerals Yearbook—1999* (Reston, VA: 1999); aluminum from "Chapter 5. Case Studies on Minerals," in MMSD, op. cit. note 10, p. 3.

## OIL SPILLS DECLINE (pages 68–69)

1. Elise DeCola, "International Oil Spill Statistics: 2000," *Oil Spill Intelligence Report* (OSIR) (Arlington, MA: Cutter Information Corp., 2001), p. 8. Included are accidental, operational, and intentional spills of oil cargo, fuel and bunker oils, and bilge oil that involve the loss of at least 34 tons (10,000 gallons) each. Releases of oil that occur slowly over a long period of time, and are therefore not captured as a spill event, are not included. The statistics refer to oil spilled without regard to any amounts that may subsequently have been recovered (which is often difficult to quantify and confirm).
2. DeCola, op. cit. note 1, pp. 26–31.
3. Ibid.
4. Ibid., p. 10.
5. Ibid., p. 12.
6. Saul Bloom et al., eds., *Hidden Casualties: Environmental, Health and Political Consequences of the Persian Gulf War* (Berkeley, CA: North Atlantic Books, 1994), p. 46.
7. DeCola, op. cit. note 1, p. 7. OSIR regards the Chechnya report as still unconfirmed.
8. Calculated from DeCola, op. cit. note 1, pp. 26–31.
9. Calculated from ibid., pp. 15–18. War-related spills have only been tracked since 1978. The OSIR database does not fully capture the total quantity of oil released due to acts of war because in some cases there is no reliable information about how much was released. The data reported here are therefore an underestimate.
10. Calculated from DeCola, op. cit. note 1, p. 13.
11. OSIR, "Oil Spill Basics: A Primer for Students," at <www.cutter.com/osir/primer.htm>, viewed 6 January 2002.

12. DeCola, op. cit. note 1, pp. 26–31; International Tanker Owners Pollution Federation, "Historical Data," at <www.itopf.com/stats.html>, viewed 30 October 2001.

13. Calculated from DeCola, op. cit. note 1, pp. 15–18, 26–31.

14. Safety measures from International Maritime Organization, at <www.imo.org>; reduced number of accidents and quantity spilled from International Tanker Owners Pollution Federation, op. cit. note 12.

15. DeCola, op. cit. note 1, pp. 13–14.

16. Ibid.

17. Ibid., p. 25.

18. Global Exchange and Essential Action, *Oil for Nothing: Multinational Corporations, Environmental Destruction, Death and Impunity in the Niger Delta* (San Francisco, CA, and Washington, DC: January 2000); deaths from "Pipeline Explodes in Nigeria," *BBC News Online,* 9 November 2001.

19. *BBC News Online,* 7 January 2002.

20. DeCola, op. cit. note 1, pp. 31–32. Because the amounts spilled could not be confirmed, they are not included in the OSIR data.

21. DeCola, op. cit. note 1, pp. 31–32.

22. Ibid., p. 13.

23. Ibid.

24. U.S. Environmental Protection Agency, "Understanding Oil Spills and Oil Spill Response," at <www.epa.gov/oilspill/docs/pdfbook.htm>; "Oil Tanker Grounded in Plymouth Sound Conservation Area," *Environment News Service,* 3 January 2002; Curtis Rist, "Survival of the Slickest," *Discover,* January 2002, p. 54.

25. Exxon Valdez Oil Spill Trustee Council, "Does Oil Remain on the Beaches 10 Years Later?" at <www.oilspill.state.ak.us/beaches/beaches.htm>, viewed 6 January 2002.

26. Greenpeace, "Exxon Valdez to Northstar: The Impacts of Oil Development in Alaska and the Arctic," at <www.greenpeace.org/~climate/arctic99/reports/exxon1.html>, viewed 14 January 2002.

## ROUNDWOOD PRODUCTION REBOUNDS (pages 70–71)

1. U.N. Food and Agriculture Organization (FAO), *FAOSTAT Statistical Database,* at <apps.fao.org>, updated 16 November 2001.

2. Ibid.

3. Ibid.

4. Ibid.

5. Ibid. Nations of the former Soviet Union and the Eastern bloc are included in industrial nations.

6. FAO, op. cit. note 1.

7. Ibid.

8. Ibid.

9. Birger Solberg et al., "An Overview of Factors Affecting the Long-Term Trends of Non-Industrial and Industrial Wood Supply and Demand," in Birger Solberg, ed., *Long-Term Trends and Prospects in World Supply and Demand for Wood and Implications for Sustainable Forest Management* (Joensuu, Finland: European Forestry Institute, 1996).

10. FAO, op. cit. note 1.

11. Ibid.

12. Ibid.

13. FAO, *State of the World's Forests 1997* (Oxford, U.K.: 1997).

14. FAO, *State of the World's Forests 2001* (Rome: 2001), p. 14. China is counted as a developing country.

15. FAO, op. cit. note 1.

16. Ibid.; International Institute for Environment and Development, *Towards a Sustainable Paper Cycle* (London: 1996).

17. FAO, op. cit. note 1.

18. Ibid.

19. Ibid.

20. Neil Scotland et al., "Indonesia Country Paper on Illegal Logging: Executive Summary and Recommendations," prepared for World Bank–WWF Workshop on Control of Illegal Logging in East Asia, Jakarta, 28 August 2000, p. 2; FAO, op. cit. note 14, p. 92.

21. Arnaldo Contreraas-Hermosilla, "Law Compliance in the Forestry Sector: An Overview," prepared for the Forest Law Enforcement and Governance East Asia Regional Ministerial Conference, Bali, Indonesia, 11–13 September 2001; Brazil also from Greenpeace International, *Partners in Mahogany Crime* (Amsterdam: October 2001); Russia also from Bureau for Regional Oriental Campaigns, Friends of the Earth-Japan, and the Pacific Environment and Resources Center, *Plundering Russia's Far Eastern Taiga: Illegal Logging, Corruption and Trade* (2000), at <www.foejapan.org>; Cambodia from Global Witness, *The Credibility Gap—And the Need to Bridge It: Increasing the Pace of Forestry Reform* (London: May 2001); Global Witness, *Tay-*

lor-made: The Pivotal Role of Liberia's Forests and Flag of Convenience in Regional Conflict (London: September 2001); Cameroon also from Henrietta Bikie et al., An Overview of Logging in Cameroon (Washington, DC: Global Forest Watch Cameroon/World Resources Institute, 2000).

22. See, for example, the Ministerial Declaration of the Forest Law Enforcement and Governance East Asia Regional Ministerial Conference, Bali, Indonesia, 11–13 September 2001, and the forthcoming Africa Forest Law Enforcement and Governance Conference.

23. Figure for 2001 from Forest Stewardship Council, "Forests Certified by FSC-Accredited Certification Bodies," 13 February 2002, at <www.fscoax.org>, viewed 3 March 2002; 1998 area from Forest Stewardship Council, "Forests Certified by FSC-Accredited Certification Bodies," August 1998, at <www.fscoax.org>, viewed 19 October 1998.

24. Forest Stewardship Council, "Forests Certified," 13 February 2002, op. cit. note 23.

## VEHICLE PRODUCTION DECLINES SLIGHTLY (pages 74–75)

1. DRI-WEFA Global Automotive Group, Global Production of Light Vehicles by Region & Country December 2001, received from Colin Couchman, e-mail to author, 8 January 2002; earlier data from Standard and Poor's DRI, World Car Industry Forecast Report, December 2000 and December 1999 (London: 2000 and 1999), and from American Automobile Manufacturers Association (AAMA), World Motor Vehicle Facts and Figures 1998 (Washington, DC: 1998).

2. DRI-WEFA Global Automotive Group, Global Sales of Light Vehicles by Region & Country December 2001, received from Colin Couchman, e-mail to author, 8 January 2002.

3. DRI-WEFA, op. cit. note 1; DRI-WEFA, op. cit. note 2.

4. Colin Couchman, DRI-WEFA Global Automotive Group, e-mail to author, 8 January 2002.

5. PricewaterhouseCoopers, "Light Vehicle Assembly by Region, Country, Category," 2001 Q4 Vehicle Outlook Reports, at <www.autofacts.com>, viewed 16 December 2001.

6. Ibid.

7. Ward's Communications, Ward's Motor Vehicle Facts & Figures 2001 (Southfield, MI: 2001), p. 61.

8. Ibid.

9. Ibid., p. 60; AAMA, Motor Vehicle Facts & Figures 1995 (Detroit, MI: 1995), p. 54. Materials consumption data are for the entire motor vehicle industry, including production of passenger cars, trucks, and buses.

10. Ward's Communications, op. cit. note 7, p. 60; AAMA, op. cit. note 9, p. 54.

11. U.S. share of world passenger cars calculated from Ward's Communications, op. cit. note 7, pp. 51, 53.

12. Ibid., p. 83.

13. Ibid.; Oak Ridge National Laboratory (ORNL), Transportation Energy Databook 21 (Oak Ridge, TN: October 2001), Table 7.16.

14. Ward's Communications, op. cit. note 7, p. 83; ORNL, op. cit. note 13, Table 7.16.

15. ORNL, op. cit. note 13, Table 1.5.

16. Lew Fulton, International Energy Agency, e-mail to author, 20 December 2001. One reason for higher fuel economy in Europe is the extensive reliance on diesel-powered cars, which burn as much as 30 percent less fuel than gasoline cars with engines of comparable size; Edmund L. Andrews with Keith Bradsher, "It Gets 78 Miles a Gallon, But U.S. Snubs Diesel," New York Times, 27 May 2001.

17. "Modest U.S. Goals on Fuel Economy," New York Times, 15 May 2001.

18. Fulton, op. cit. note 16.

19. Steven Plotkin, Argonne National Laboratory, "European and Japanese Fuel Economy Initiatives: What They Are, Their Prospects for Success, Their Usefulness as a Guide for U.S. Action," Energy Policy, November 2001, pp. 1073–84; "Modest U.S. Goals on Fuel Economy," op. cit. note 17.

20. Plotkin, op. cit. note 19; "Modest U.S. Goals on Fuel Economy," op. cit. note 17.

21. Julie Vorman, "U.S. Science Panel Says Detroit Can Improve Fuel Use," Reuters, 17 January 2002.

22. Plotkin, op. cit. note 19; "Modest U.S. Goals on Fuel Economy," op. cit. note 17.

23. "Panel Says 80-mpg Sedan Not Likely by 2004," Reuters, 14 August 2001; Neela Banerjee with Danny Hakim, "U.S. Ends Car Plan on Gas Efficiency; Looks to Hydrogen," New York Times, 9 January 2002.

24. "Weststart Speaks from a Decade of Experience in Alternative Vehicles," Environmental Business Journal, vol. 13, no. 7/8 (2001), p. 4.

# Notes

## BICYCLE PRODUCTION ROLLS FORWARD
(pages 76–77)

1. Based on "World Market Report 2002," in Bicycle Retailer and Industry News, *Industry Directory 2002* (Santa Fe, NM: Bill Communications, 2002), and on United Nations, *Industrial Commodity Statistics Yearbook 1999* (New York: 2000).
2. Worldwatch calculation based on "World Market Report 2002," op. cit. note 1, and on United Nations, op. cit. note 1.
3. "World Market Report 2002," op. cit. note 1.
4. Worldwatch calculation based on "World Market Report 2002," op. cit. note 1, and on United Nations, op. cit. note 1.
5. "World Market Report 2002," op. cit. note 1.
6. Ibid.
7. Ibid.
8. Ibid.
9. Ibid.
10. Worldwatch calculation based on "World Market Report 2002," op. cit. note 1, and on United Nations, op. cit. note 1.
11. Worldwatch calculation based on "World Market Report 2002," op. cit. note 1, and on United Nations, op. cit. note 1.
12. Institute for Transportation and Development Policy, "TransportActions," at <www.itdp.org/>, viewed 12 February 2002; Oscar Edmundo Diaz, Institute for Transportation and Development Policy, discussion with author, 14 February 2002.
13. Pierre Graftieaux, World Bank, discussion with author, 13 February 2002.
14. Ibid.
15. Federal Highway Administration, *Case Study No. 1: Reasons Why Bicycling and Walking are Not Being Used More Extensively as Travel Modes* (Washington, DC: U.S. Department of Transportation, 1992), p. 20.
16. John Pucher and Lewis Dijkstra, "Making Walking and Cycling Safer: Lessons from Europe," *Transportation Quarterly*, vol. 54, no. 3 (2000), pp. 25–50.
17. Ibid.
18. Ed Benjamin and Frank Jamerson, bicycle industry consultants, e-mail to author, 8 February 2002.
19. "Experts Offer Conflicting Views on E-bike Future," *Bicycle Retailer and Industry News*, 1 May 2001.
20. "Inventions of the Year: 2001," at <www.time.com>, 11 November 2001, viewed 21 January 2002; Lucy Chubb, "New Bicycle Gets Big Push from Fuel Cells," *Environmental News Network*, viewed 3 January 2002.
21. "Inventions of the Year," op. cit. note 20; Chubb, op. cit. note 20.
22. "Inventions of the Year," op. cit. note 20; Chubb, op. cit. note 20.

## PASSENGER RAIL AT CROSSROADS
(pages 78–79)

1. World Bank, "Railways Database," at <www.worldbank.org/html/fpd/transport/rail/rdb.htm>, viewed 12 December 2001; Louis Thompson, Railways Advisor, World Bank, e-mail to author, 10 January 2002.
2. Road travel from International Road Federation, *World Road Statistics* (Geneva: various editions); air travel from Attilio Costaguta, Chief, Statistics and Economic Analysis Section, International Civil Aviation Organization, Montreal, e-mail to Lisa Mastny, Worldwatch Institute, 2 November 1998.
3. Louis S. Thompson and Julie M. Fraser, "World Bank Railway Database," *Infrastructure Notes, Transport No. RW-6* (Washington, DC: Transportation, Water and Urban Development Department, World Bank, October 1993); Figure 2 from World Bank, op. cit. note 1, and from Thompson, op. cit. note 1.
4. Ibid.
5. Arnulf Grübler, *The Rise and Fall of Infrastructures: Dynamics of Evolution and Technological Change in Transport* (New York: Springer-Verlag, 1990); Arnulf Grübler and Nebojsa Nakićenović, *Evolution of Transport Systems: Past and Future* (Laxenburg, Austria: International Institute for Applied Systems Analysis, 1991).
6. Competitiveness at medium distances from World Bank, "Railways Overview," at <www.worldbank.org/html/fpd/transport/rail_ss.htm>, viewed 12 December 2001; energy intensities from Stacy C. Davis, ed., *Transportation Energy Data Book, Edition 21* (Oak Ridge, TN: U.S. Department of Energy, Oak Ridge National Laboratory, 2001).
7. Jean-Claude Raoul, "How High-Speed Trains Make Tracks," *Scientific American*, October 1997, pp. 100–05.
8. Ibid.

9. "European Trains Just Keep Getting Faster and Faster," *Christian Science Monitor*, 28 March 2001.

10. Mike Knutton, "Air/Rail Collaboration: Heaven or Hell?" *Railway Age*, September 2001, p. 88; James P. RePass, "Planes to the Trains Coming to America?" *Railway Age*, January 2001.

11. John Tagliabue, "Airlines Feel Pressure of Europe's Fast Trains," *New York Times*, 12 August 2001.

12. Thalys from ibid.

13. Shen Bin, "China to Double Rail Spending," *China Daily*, 30 March 1998.

14. Li Rongxia, "Chinese Trains Run Faster," *Beijing Review*, 27 November 2000.

15. Chen Qide, "High-Speed Railway Line Construction Scheduled," *China Daily*, 1 April 1998.

16. Cecilia M. Kang, "Back on Track: Seoul Gives the Green Light to Much-Needed Rail System," *Far Eastern Economic Review*, 23 April 1998; Laura Tyson, "Taiwan's High-Speed Rail Project is Right on Track," *Financial Times*, 18 March 1998.

17. Louis S. Thompson, "The Benefits of Separating Rail Infrastructure from Operations," *Viewpoint* (World Bank Group Finance, Private Sector, and Infrastructure Network), Note No. 135, December 1997.

18. Anthony Perl and James A. Dunn, Jr., "Fast Trains: Why the U.S. Lags," *Scientific American*, October 1997, pp. 106–08.

19. T. R. Reid, "Finger-Pointing Begins in Crash of British Trains; Privatized System Cited As Source of Problems," *Washington Post*, 18 October 1999; Sarah Lyall, "Railways' Frightful State is the Talk of Britain," *New York Times*, 10 December 2000; "Unfare Treatment," *Railwatch*, December 1997, p. 1; Juliette Jowit, "National User Group Defends Plans for Rail," *Financial Times*, 15 October 2001.

20. David Pringle and Masayoshi Kanabayashi, "Experience of British Rail System Serves as Cautionary Tale—Germany Amends Plan In Effort to Avoid Problems of Railtrack," *Wall Street Journal*, 17 October 2001.

21. Thompson, op. cit. note 1.

22. Service needs from World Bank, op. cit. note 6; global standards from Ken Harris, ed., *Jane's World Railways 1999–2000* (London: 1999); technology from Tracy Staedter, "Magnetically Levitated Trains," *Technology Review*, October 2001, pp. 86–87.

## INTERNET CONTINUES METEORIC RISE
(pages 82–83)

1. Host computer count from Internet Software Consortium, "Internet Domain Surveys 2000–02," at <www.isc.org/ds/>, and from Network Wizards, "Internet Domain Surveys, 1981–1999," at <www.nw.com>, viewed 20 February 2002. A single host computer can wire several computers to the Internet in the same way that one telephone line can plug in multiple phone extensions. Number of users is a Worldwatch estimate based on Nua, Ltd., "How Many Online?" at <www.nua.ie>, updated February 2002, and on China Internet Network Information Center (CINIC), "SemiAnnual Survey Report on the Development of China's Internet (January 2002)," at <www.cnnic.net.cn/develst/rep200201-e. shtml>, viewed 1 March 2002. Users are estimated in terms of individuals who use the Internet on a weekly basis. Because estimates for the number of users can vary, host computers provide a more reliable measure of the Internet's size.

2. Internet Software Consortium, op. cit. note 1; Network Wizards, op. cit. note 1.

3. Based on Nua, Ltd., op. cit. note 1.

4. Ibid.

5. Ibid.

6. Based on CINIC, op. cit. note 1, and on Nua, Ltd., op. cit. note 1.

7. Based on CINIC, op. cit. note 1, and Ward's Communications, *World Motor Vehicle Data 2000* (Southfield, MI: 2000), p. 14.

8. Based on Nua, Ltd., op. cit. note 1.

9. Based on ibid., and on Population Reference Bureau (PRB), *World Population Datasheet*, wallchart (Washington, DC: 2001).

10. Ibid.

11. Ibid.

12. Ibid.; CINIC, op. cit. note 1.

13. Based on Nua Ltd., op. cit. note 1, and on PRB, op. cit. note 9.

14. Worldwatch estimate based on Nua, Ltd., op. cit. note 1, on CINIC, op. cit. note 1, and on PRB, op. cit. note 9.

15. Nua, Ltd., op. cit. note 1.

16. Ibid.

17. Ibid.

18. Global Reach, "Global Internet Statistics (by Language)," at <www.glreach.com/globstats/index.

# Notes

php3>, updated 15 December 2001.

19. Ibid.

20. Frances Williams, "Chinese to Become Most-Used Language on Web," *Financial Times*, 7 December 2001.

21. Eileen X. Grant, "Study: E-Commerce to Top $1 Trillion in 2002," *E-Commerce Times*, 13 February 2002.

22. U.S. figure from ibid.; Japan from International Telecommunications Union (ITU), "Internet and Electronic Commerce," news release (Geneva: 17 May 2001).

23. Mercedes Cardona, "2001 Ad Spending Worse than Predicted," *AdAge.com*, 5 March 2002; CyberAtlas, "Internet Will Lead Advertising Sector Back," 24 January 2002, at <cyberatlas.internet.com/markets/ advertising>.

24. John Schwartz, "Page by Page History of the Web," *New York Times*, 29 October 2001.

25. Current figure from ibid.; 1998 from Alexa Internet, "Web Spawns 1.5 Million Pages Daily According to Findings from Alexa Internet," press release (San Francisco, CA: 31 August 1998).

26. ITU, "Internet and Health: Is there a Doctor?" news release (Geneva: 17 May 2001).

27. ITU, "The Internet and Education," news release (Geneva: 17 May 2001).

28. Basel Action Network and Silicon Valley Toxics Coalition, *Exporting Harm: The High-Tech Trashing of Asia* (Seattle, WA, and San Jose, CA: February 2002).

29. Ibid.

## MOBILE PHONE USE BOOMS (pages 84–85)

1. Data for 2001 from Tim Kelly, International Telecommunication Union (ITU), e-mail to author, 5 March 2002, and from ITU, "Mobile Phone Market Penetration," press release (Geneva: 8 February 2002); 2000 and 1999 from idem, "Cellular Subscribers," 9 January 2002, at Free Statistics Homepage, <www.itu.int/ITU-D/ict/statistics/>, viewed 4 February 2002; 1960–98 from idem, *Telecommunication Indicators Update,* Socioeconomic Time-series Access and Retrieval System database, viewed 24 August 1999.

2. ITU, "Mobile Phone Market," op. cit. note 1; idem, "Cellular Subscribers," op. cit. note 1; idem, *Telecommunication Indicators Update*, op. cit. note 1.

3. ITU, "Mobile Phone Market," op. cit. note 1; idem,

"Cellular Subscribers," op. cit. note 1; idem, *Telecommunication Indicators Update*, op. cit. note 1.

4. Thomas B. Allen, "The Future is Calling," *National Geographic*, December 2001, pp. 76–83.

5. ITU, "Mobile Phone Market," op. cit. note 1.

6. ITU, "Cellular Subscribers," op. cit. note 1.

7. Alan Cowell, "Not in Finland Anymore? More Like Nokialand," *New York Times*, 6 February 2002.

8. "Gartner Dataquest Reports Worldwide Mobile Phone Shipments Declined 10 Percent in the Third Quarter of 2001," press release, 19 November 2001, at <www.semiseeknews.com/press_release2787.htm>, viewed 4 February 2002.

9. ITU, "Cellular Subscribers," op. cit. note 1.

10. ITU, "Mobile Phone Market," op. cit. note 1.

11. ITU, "Cellular Subscribers," op. cit. note 1.

12. ITU, *Telecommunication Indicators Update*, October-November-December 2000 (Geneva: 2000).

13. ITU, "Mobile Phone Market," op. cit. note 1.

14. ITU, *Telecommunication Indicators Update*, April-May–June 2000 (Geneva: 2000).

15. Mark Turner, "The Call of Africa Grows Louder," *Financial Times*, 21 August 2001.

16. James Lamont, "Africans Get Connected to Mobile Telephony," *Financial Times*, 12 November 2001.

17. ITU, *Telecommunication Indicators Update*, July-August-September-October 2001 (Geneva: 2001).

18. ITU, "Cellular Subscribers," op. cit. note 1.

19. Fiona Harvey, "Low-Tech Entry to a High-Tech World," *Financial Times*, 5 December 2001.

20. Muhammad Yunus, "Microcredit and IT for the Poor," *New Perspectives Quarterly*, winter 2001, pp. 25–26.

21. Ibid.; Farid Ahmed and Abou Abel, "Hello, I'm Calling from Parulia..." *The UNESCO Courier*, July/August 2000, pp. 67–68.

22. Reuters, "Long Road to Phone-Car Laws," *Wired News*, 9 May 2001.

23. Karen Solomon, "When Toxic Waste Comes Calling," *Wired News*, 8 March 2001.

24. David Edwards, "Hold That Call," *The Ecologist*, October 2001, pp. 26–28.

25. Gautam Malkani, "Mobile Phone Safety Probed by 15 Studies," *Financial Times*, 26 January 2002.

## POPULATION GROWING STEADILY
(pages 88–89)

1. U.S. Bureau of the Census, *International Data Base*, electronic database, Suitland, MD, updated 10 May 2000.
2. Ibid.
3. Ibid.
4. Population Reference Bureau (PRB), *2001 World Population Data Sheet*, wall chart (Washington, DC: 2001).
5. United Nations, *World Population Prospects: The 2000 Revision* (New York: 2001).
6. PRB, op. cit. note 4.
7. Ibid.
8. United Nations, op. cit. note 5.
9. Peter G. Peterson, "Gray Dawn: The Global Aging Crisis," *Foreign Affairs*, January/February 1999, pp. 42–55; Phillip J. Longman, "The Global Aging Crisis," *U.S News & World Report*, 1 March 1999, pp. 30–39.
10. Census Bureau, op. cit. note 1.
11. Ibid.
12. United Nations, op. cit. note 5.
13. Ibid.
14. Martin Brockerhoff and Ellen Brennan, *The Poverty of Cities in the Developing World*, Policy Research Dvision Working Paper No. 96 (New York: Population Council, 1997), p. 5.
15. Estimate of developing-country contraceptive prevalence in the late 1960s from J. Khanna, P. F. A. Van Look, and P. D. Griffin, *Reproductive Health: Key to a Brighter Future, World Health Organization Biennial Report 1990–1991* (Geneva: World Health Organization, 1992), pp. 5–6; modern contraceptive prevalence from PRB, op. cit. note 4.
16. Hantamalala Rafalimanana and Charles F. Westoff, "Potential Effects on Fertility and Child Health Survival of Birth-Spacing Preferences in Sub-Saharan Africa," *Studies in Family Planning*, June 2000, p. 99.
17. "Ending Violence Against Women," *Population Reports*, December 1999, p. 14.
18. John A. Ross and William L. Winfrey, "Unmet Need in the Developing World and the Former USSR: An Updated Estimate," unpublished manuscript, received 1 November 2001.
19. U.N. Population Fund (UNFPA), *The State of World Population 1999* (New York: 1999), p. 29.
20. Population Action International, "What You Need to Know About the Global Gag Rule: An Unofficial Guide" (Washington, DC: August 2001).
21. United Nations, op. cit. note 5.
22. UNFPA, *The State of the World Population 2000* (New York: 2000), p. 23.
23. UNFPA, *The State of the World Population 2001* (New York: 2001), p. 41; United Nations, *The World's Women 2000: Trends and Statistics* (New York: 2000), pp. 87, 91.
24. United Nations, op. cit. note 23, pp. 164–67.

## AIDS PASSES 20-YEAR MARK (pages 90–91)

1. Joint United Nations Programme on HIV/AIDS (UNAIDS), *AIDS Epidemic Update: December 2001* (Geneva: 2001), p. 1; UNAIDS, *AIDS Epidemic Update: December 2000* (Geneva: 2000); Neff Walker, UNAIDS, e-mail to Brian Halweil, Worldwatch Institute, 20 March 2000; population of Afghanistan from U.S. Bureau of the Census, *International Data Base*, electronic database, Suitland MD, updated 10 May 2000.
2. UNAIDS, *December 2001*, op. cit. note 1, p. 1.
3. Ibid., p. 3.
4. Christopher J. L. Murray et al., *The Global Burden of Disease 2000 Project: Aims, Methods, and Data Sources*, Global Programme on Evidence for Health Policy Discussion Paper No. 36 (Geneva: World Health Organization, November 2001), p. 18.
5. UNAIDS, *December 2001*, op. cit. note 1, p. 8.
6. Ibid., p. 7.
7. Ibid., p. 10.
8. Monitoring the AIDS Pandemic Network, *The Status and Trends of HIV in Asia and the Pacific: October 2001* (Melbourne, Australia: 2001).
9. Elisabeth Rosenthal, "AIDS Patients in China Lack Effective Treatment," *New York Times*, 12 November 2001.
10. R. D'Amelio et al., "A Global Review of Legislation on HIV/AIDS: The Issue of HIV Testing," *Journal of Acquired Immune Deficiency Syndromes*, vol. 28, no. 2 (2001), pp. 173–79.
11. Avert, "The History of Aids 1981–1986," at <www.avert.org/his81_86.htm>, viewed 13 January 2002.
12. David Brown, "Study Finds Drug-Resistant HIV in Half of Infected Patients," *Washington Post*, 19 December 2001.
13. UNAIDS, "Accelerating Access to HIV Care, Support and Treatment," at <www.unaids.org/acc_

# Notes

access/index.html>, viewed 13 January 2002.

14. "Africa's $4.5 Billion AIDS Underspend," *CNN.com,* 11 December 2001.

15. Dina Kraft, "Court Orders Government to Provide Key AIDS Drug to Pregnant Women," *Associated Press,* 14 December 2001.

16. Sheryl Gay Stolberg, "AIDS Fund Falls Short of Goal and U.S. Is Given Some Blame," *New York Times,* 13 February 2002.

17. Laurie Garrett, "A Push to Cut Cost of AIDS Drugs," *Newsday,* 9 November 2001.

## NUMBER OF VIOLENT CONFLICTS DECLINES (pages 94–95)

1. Number of wars from Arbeitsgemeinschaft Kriegsursachenforschung (AKUF), "Im Schatten des Anti-Terror-Kriegs: Weltweit 46 Kriegerische Konflikte im Jahr 2001," press release (Hamburg, Germany: Institute for Political Science, University of Hamburg), 17 December 2001, and from Wolfgang Schreiber, AKUF, e-mail to author, 19 December 2001; historical data from ibid., from Klaus Jürgen Gantzel and Torsten Schwinghammer, *Die Kriege nach dem Zweiten Weltkrieg 1945 bis 1992* (Münster and Hamburg: Lit Verlag, 1995), from Dietrich Jung, Klaus Schlichte, and Jens Siegelberg, *Das Kriegsgeschehen 1995* (Bonn: Stiftung Entwicklung und Frieden, 1996), from Thomas Rabehl, AKUF, e-mail to author, 8 January 1998, from AKUF, "Kriegesbilanz 1998," press release, 11 January 1999, from AKUF, "Weltweit 35 Kriege in 1999," press release, 20 December 1999, from Wolfgang Schreiber, AKUF, e-mail to author, 21 January 2001, and from AKUF, "Das Kriegsgeschehen des Jahres 2000," press release, December 2000. AKUF defines a "war" as a mass conflict in which regular armed forces are involved on at least one side, where there is at least a minimal degree of central organization, and where there is a continuity of military operations rather than a series of spontaneous, occasional clashes. "Armed conflicts" are those cases where its definition of "war," particularly the criterion of a continuity of armed operations, is not fully met or where a lack of information does not justify including a violent conflict in the "war" category.

2. AKUF, "Im Schatten des Anti-Terror-Kriegs," op. cit. note 1.

3. Ibid.

4. Ibid.

5. Ted Robert Gurr, Monty G. Marshall, and Deepa Khosla, *Peace and Conflict 2001* (College Park, MD: Center for International Development and Conflict Management, University of Maryland, 2001), pp. 7–8.

6. Ibid., p. 9.

7. AKUF, "Im Schatten des Anti-Terror-Kriegs," op. cit. note 1.

8. Gurr, Marshall, and Khosla, op. cit. note 5, p. 2.

9. For a discussion of how resource plunder drives violent conflict, see Michael Renner, "Breaking the Link Between Resources and Repression," in Worldwatch Institute, *State of the World 2002* (New York: W.W. Norton & Company, 2002), pp. 149–73.

10. Margareta Sollenberg, ed., *States in Armed Conflict 2000*, Report No. 60 (Uppsala, Sweden: Uppsala University, Department of Peace and Conflict Research, 2001), p. 10; Nils Petter Gleditsch et al., "Armed Conflict 1946–99: A New Dataset," prepared for Identifying Wars: Systematic Conflict Research and its Utility in Conflict Resolution and Prevention, Conference on Data Collection in Armed Conflict, Uppsala, Sweden, 8–9 June 2001.

11. Sollenberg, op. cit. note 10, p. 10.

12. Ibid.

13. Heidelberger Institut für Internationale Konfliktforschung (HIIK), *Konfliktbarometer 2001* (Heidelberg, Germany: Institute for Political Science, University of Heidelberg, 2001), p. 3, and earlier editions of this report.

14. Ibid.

15. Ibid., p. 6.

16. Milton Leitenberg, "Deaths in Wars and Conflicts Between 1945 and 2000," May 2001, Center for International and Security Studies, University of Maryland, paper prepared for Conference on Data Collection in Armed Conflict, Uppsala, Sweden, 8–9 June 2001. Leitenberg includes civilian and military deaths, combatant and noncombatant casualties, and deaths as a result not only of direct violence but also conflict-induced disease and starvation.

17. Leitenberg, op. cit. note 16.

## PEACEKEEPING EXPENDITURES RISE AGAIN (pages 96–97)

1. U.N. Department of Public Information (UNDPI), "United Nations Peacekeeping Operations. Back-

ground Note," New York, 15 January 2002. (Beginning with July 1997, the United Nations switched its peacekeeping accounts from calendar years to July–June reporting periods.) Expenditure figure also based on Luisa Anzola, U.N. Department of Peacekeeping Operations (UNDPKO), Field Administration and Logistics Division, New York, discussion with author, 20 December 1995, on "Peace-Keeping Operations Expenditures (All Missions)," at <www.un.org/Depts/DPKO/yir96/allexp.jpg>, viewed 18 March 1997, on U.N. General Assembly, "Support for Peacekeeping Operations," A/52/837, New York, 20 March 1998, p. 12, on UNDPI, "Background Note: U.N. Peacekeeping Operations," New York, 15 June 2000, and on author's computations.

2. UNDPKO, "Monthly Summary of Contributors," at <www.un.org/Depts/dpko/dpko/contributors/index.htm>, viewed 28 January 2002. Personnel number also based on William Durch, Henry Stimson Center, Washington, DC, e-mail to author, 9 January 1996, and on Global Policy Forum, at <www.globalpolicy.org/security/peace kpg/data/pkomctab.htm>, viewed 28 January 2002.

3. As of 31 December 2001; UNDPI, "Background Note," 15 January 2002, op. cit. note 1.

4. UNDPI, "United Nations Political and Peace-Building Missions. Background Note," New York, 1 June 2001.

5. UNDPI, "Background Note," 15 January 2002, op. cit. note 1.

6. Global Policy Forum, "Troop and Other Personnel Contributions to Peacekeeping Operations: 2001," at <www.globalpolicy.org/security/peace kpg/data/pktp01.htm>, viewed 28 January 2002.

7. Author's calculation, based on data from Global Policy Forum, op. cit. note 6, and on UNDPKO, op. cit. note 2.

8. Global Policy Forum, op. cit. note 6; UNDPKO, op. cit. note 2.

9. Global Policy Forum, op. cit. note 6; UNDPKO, op. cit. note 2.

10. Barbara Crossette, "U.S. Vetoes U.N. Council Bid on Palestinian Force," New York Times, 29 March 2001; United Nations, "Security Council Fails to Adopt Resolution on Middle East Situation, to Condemn Use of Force, Encourage Monitoring Mechanism," press release (New York: 15 December 2001).

11. Barbara Crossette, "U.N. Rejects Troops for Palestinian Territories," New York Times, 19 December 2000.

12. UNDPI, "Background Note," 15 January 2002, op. cit. note 1; UNDPKO, "Current Peacekeeping Operations," at <www.un.org/Depts/DPKO/c_miss.htm>, viewed 10 December 2001.

13. UNDPI, "Background Note," 15 January 2002, op. cit. note 1.

14. Ibid.

15. Ibid.

16. Ibid.

17. Money owed by all members was $1.919 billion ($1.886 billion in 2000 dollars); "Status of Contributions to the Regular Budget, International Tribunals and Peacekeeping Operations as at 31 December 2001," United Nations, Office of the Spokesman for the Secretary-General, New York.

18. Ibid.

19. Lizette Alvarez, "House Approves $582 Million For Back Dues Owed to U.N.," New York Times, 25 September 2001.

20. U.N. General Assembly, "After Several Years in Red, Financial Stability for UN 'Close at Hand'," GA/AB/3456, New York, 10 October 2001.

21. Compiled from Thomas Papworth, "Multilateral Peace Operations, 2000," in Stockholm International Peace Research Institute, SIPRI Yearbook 2001. Armaments, Disarmament and International Security (New York: Oxford University Press, 2001), pp. 128–48, and from International Institute for Strategic Studies, "The 2001 Chart of Armed Conflict," wall chart distributed with The Military Balance 2001–2002 (London: Oxford University Press, 2001).

22. United Nations, "Security Council Resolutions," at <www.un.org/Docs/scres/2001/sc2001.htm>, 20 December 2001; "Afghanistan: U.S. To Have Authority Over Peacekeepers; More," UN Wire, 20 December 2001.

## FARMLAND QUALITY DETERIORATING
(pages 102–03)

1. Sara J. Scherr, "Productivity-Related Economic Impacts of Soil Degradation in Developing Countries: An Evaluation of Regional Experience," University of Maryland, College Park, manuscript sent to author, 13 December 2001.

2. World's cropland area from Stanley Wood et al., Pilot

# Notes

*Analysis of Global Ecosystems: Agroecosystems* (Washington, DC: World Resources Institute, 2000), p. 21; percent of land affected by degradation is author's calculation based on table in ibid., p. 48. Wood et al. note that most coverage of their findings misinterpreted the finding that "over 40 percent of the agricultural extent coincides with mapping units whose degradation severity is high or very high" to mean that over 40 percent of the world's agricultural lands suffer from high or very high degradation. This overstates the problem because it assumes that the whole mapping unit suffers from the same severity of degradation. Often, just a portion of the mapping unit is degraded.

3. Percent of land affected by degradation is author's calculation based on Wood et al., op. cit. note 2, p. 48.

4. L. R. Oldeman, "The Global Extent of Land Degradation," in D. J. Greenland and I. Szabolcs, eds., *Land Resilience and Sustainable Land Use* (Wallingford, Oxford, U.K.: CABI, 1994), pp. 99–118.

5. Sara J. Scherr, "Soil Degradation: A Threat to Developing-Country Food Security by 2020?" Food, Agriculture, and the Environment Discussion Paper 27 (Washington, DC: International Food Policy Research Institute (IFPRI)), February 1999), p. 16.

6. Scherr, op. cit. note 1, p. 2.

7. A. T. Ayoub, "Indicators of Dryland Degradation," in Victor R. Squires and Ahmed E. Sidahmed, eds., *Drylands: Sustainable Use of Rangelands into the Twenty-First Century* (Rome: International Fund for Agricultural Development, 1998), p. 12.

8. Over 70 percent from ibid., p. 13; 3.4 billion hectares from U.N. Food and Agriculture Organization (FAO), *FAOSTAT Statistical Database*, at <apps.fao.org>, updated 10 July 2001.

9. International Soil Reference and Information Centre (ISRIC), *Global Assessment of the Status of Human-Induced Soil Degradation* (GLASOD), at <www.isric.nl/GLASOD.htm>; Wood et al., op. cit. note 2, p. 48.

10. World Bank, *From Indices to Policy Implications: Land Use in Central America* (Washington, DC: November 2000), pp. 2–3.

11. Julio Henao and Carlos Baanante, *Nutrient Depletion in the Agricultural Soils of Africa* (Washington, DC: IFPRI, October 1999).

12. Wood et al., op. cit. note 2, p. 52.

13. Table 1 from the following: J. Seghal and I. P. Abrol, *Soil Degradation in India: Status and Impact* (New Delhi: Oxford and IBH, 1994), cited in Scherr, op. cit. note 1, p. 24; M. Ali and D. Byerlee, "Productivity Growth and Resource Degradation in the Irrigated Rice Bowl of Pakistan's Punjab," in M. Bridges et al., eds., *Response to Land Degradation* (Enfield, NH: Science Publishers, 2001), pp. 186–99; S. Mantel and V. W. P. van Engelen, *The Impact of Land Degradation on Food Productivity: Case Studies of Uruguay, Argentina, and Kenya* (Wageningen, Netherlands: ISRIC, 1997), pp. 39–40; Pay Dreschel et al., "Population Density, Soil Nutrient Depletion, and Economic Growth in Sub-Saharan Africa," *Ecological Economics*, August 2001, p. 255–56; A. Young, "Land Degradation in South Asia: Its Severity, Causes and Effects Upon the People," final report for the Economic and Social Council of the United Nations, FAO, U.N. Development Programme, and U.N. Environment Programme, Rome, 1993; M. Rahman et al., "Land Degradation in Bangladesh," in Bridges et al., op. cit. this note, pp. 117–29; P. K. Joshi and D. Jha, "Farm-level Effects of Soil Degradation in Sharda Sahayak Irrigation Project," Working Papers on Future Growth in Indian Agriculture, No. 1 (Karnal, India: Central Soil Salinity Research Institute, Indian Council of Agricultural Research and IFPRI, 1991); sedimentation costs from T. Enters, "Valuing the Off-Site Effects of Land Degradation," in Bridges et al., op. cit. this note, p. 203; He Sheng, "Sandstorms Sound Ecological Alarm," *China Daily*, 16 May 2000; Caspar Henderson, "Dust Clouds of Death That Are Killing Coral," *Financial Times*, 25–26 November 2000; U.S. benefits from Roger Claassen et al., *Agri-Environmental Policy at the Crossroads: Guideposts on a Changing Landscape* (Washington, DC: Economic Research Service, U.S. Department of Agriculture (USDA), January 2001), pp. 17–18.

14. Scherr, op. cit. note 1, p. 11.

15. L. R. Oldeman, *Soil Degradation: A Threat to Food Security?* (Wageningen, Netherlands: ISRIC, 1998), p. 4.

16. Scherr, op. cit. note 5, p. 28.

17. ISRIC, op. cit. note 9.

18. Ibid.

19. USDA, National Resources Conservation Service, *Summary Report: 1997 National Resources Invento-*

ry (Washington, DC: rev. December 2000), p. 7.

20. Ibid.; based on total grain production of 323 million tons in 2001 from USDA, *Production, Supply, and Distribution*, electronic database, Washington, DC, updated December 2001.

21. Area affected by salinization from Sandra Postel, *Pillar of Sand* (New York: W.W. Norton & Company, 1999), p. 93; economic losses from ibid., p. 92.

22. Enters, op. cit. note 13, pp. 201–05.

23. S. Pagiola, "The Global Environmental Impacts of Agricultural Land Degradation in Developing Countries," in Bridges et al., op. cit. note 13, pp. 207–19.

24. Secretariat of the U.N. Convention to Combat Desertification, "Governments Seek Breakthrough for Anti-desertification Struggle," press release (Geneva: October 2001).

25. Jules Pretty and Rachel Hine, *Reducing Food Poverty with Sustainable Ag: A Summary of New Evidence*, Executive Summary (Colchester, U.K.: SAFE-World Research Project, University of Essex, February 2001), pp. 66–68.

26. Spread in North America from Carmen Sandretto, "Conservation Tillage Firmly Planted in U.S. Agriculture," *Agricultural Outlook*, Economic Research Service, USDA, Washington, DC, March 2001, pp. 5–6; spread in Asia from "Conservation Agriculture Called Next Green Revolution," *Environmental News Service*, 3 October 2001; spread in Latin America from Pretty and Hine, op. cit. note 25, pp. 66–68.

27. Pretty and Hine, op. cit. note 25, pp. 66–68.

28. Ademir Calegari, "The Spread and Benefits of No-Till Agriculture in Paraná State, Brazil," in Norman Uphoff et al., eds., *Agroecological Innovations: Increasing Food Production with Participatory Development* (London: Earthscan, forthcoming).

## FOREST LOSS UNCHECKED (pages 104–05)

1. U.N. Food and Agriculture Organization (FAO), *State of the World's Forests 2001* (Rome: 2001), p. 45.

2. FAO, *Global Forest Resources Assessment 2000* (Rome: 2002), p. xxiii.

3. FAO, op. cit. note 1, p. 44; idem, *Global Forest Resources Assessment 2000* (Rome: 2002), p. 9. When the expansion of natural forests (mostly in non-tropical areas) is accounted for, there was a net loss of 125 million hectares recorded.

4. Henrietta Bikie et al., *An Overview of Logging in Cameroon* (Washington, DC: Global Forest Watch Cameroon/World Resources Institute, 2000), p. 2; Dirk Bryant et al., *The Last Frontier Forests: Ecosystems and Economies on the Edge* (Washington, DC: World Resources Institute, 1997).

5. FAO, op. cit. note 1, p. 154–57.

6. Ibid.

7. Ibid.

8. Ibid., p. 33.

9. Ibid., p. 36.

10. Ibid.

11. Ibid., p. 38.

12. Ibid.

13. Ibid.

14. Ibid., p. 44.

15. Ibid., p. 39.

16. Ibid.

17. Ibid., p. 41.

18. Ibid., p. 40.

19. Ibid.

20. Ibid.

21. Ibid., pp. 154–57. Note that this tally includes only those nations that had at least 100,000 hectares of forest cover in 1990.

22. Ibid.

23. Ibid., pp. 154–59; Global Witness, *Taylor-made: The Pivotal Role of Liberia's Forests and Flag of Convenience in Regional Conflict* (London: September 2001).

24. FAO, op. cit. note 1, pp. 154–59.

25. Janet N. Abramovitz, *Unnatural Disasters*, Worldwatch Paper 158 (Washington, DC: Worldwatch Institute, October 2001).

26. FAO, op. cit. note 1, pp. 154–59.

27. Ibid., p. 43.

28. FAO, op. cit. note 2, pp. xxvii, 11. For example, the global forest cover estimate for 2000 is 400 million hectares larger than the estimate made in 1995 because the minimum canopy cover needed for inclusion was lowered from 20 to 10 percent in industrial countries to make it consistent with the 10-percent figure long used in developing nations. This resulted in areas previously classified as "other woodlands" being counted as "forests" for the first time.

29. FAO, op. cit. note 1, p. 31.

30. See Global Forest Watch, at <www.globalforestwatch.org>.

31. Forest Watch Indonesia and Global Forest Watch,

# Notes

*The State of the Forest: Indonesia* (Bogor, Indonesia and Washington, DC: 2002), p. xi.

32. Global Witness, op. cit. note 23; idem, *The Credibility Gap—and the Need to Bridge It: Increasing the Pace of Forestry Reform* (London: May 2001); Greenpeace, *Partners in Mahogany Crime* (London: 2001); Environmental Investigation Agency and Telepak Indonesia, *Timber Trafficking: Illegal Logging in Indonesia, Southeast Asia, and International Consumption of Illegally Sourced Timber* (London: 2001).

33. U.N. Environment Programme, *Assessment of the Status of the World's Remaining Closed Forests* (Nairobi: 2001), p. 7.

34. Ibid.

35. Ibid., p. 9; other assessments include Janet N. Abramovitz, *Taking a Stand: Cultivating a New Relationship with the World's Forests*, Worldwatch Paper 140 (Washington, DC: Worldwatch Institute, April 1998), and Bryant et al., op. cit. note 4.

## FRESHWATER SPECIES AT INCREASING RISK (pages 106–07)

1. Peter B. Moyle and Robert A. Leidy, "Loss of Biodiversity in Aquatic Ecosystems: Evidence from Fish Faunas," in P. L. Fiedler and S. K. Jain, eds., *Conservation Biology: The Theory and Practice of Nature Conservation, Preservation, and Management* (New York: Chapman and Hall, 1992).

2. Carmen Revenga et al., *Pilot Analysis of Global Ecosystems: Freshwater Ecosystems* (Washington, DC: World Resources Institute, 2000), p. 49.

3. Anthony Ricciardi and Joseph B. Rasmussen, "Extinction Rates of North American Freshwater Fauna," *Conservation Biology*, October 1999, pp. 1220–22.

4. Ibid.

5. Ibid.

6. Ibid.

7. Bruce Stein, Lynn S. Kutner, and Jonathan S. Adams, eds., *Precious Heritage: The Status of Biodiversity in the United States* (New York: Oxford University Press, 2000).

8. Ibid.

9. Ibid.

10. Table 1 from Ibid.

11. Ibid.

12. Mats Dynesius and Christer Nilsson, "Fragmentation and Flow Regulation of River Systems in the Northern Third of the World," *Science*, 4 November 1994, pp. 753–62.

13. World Commission on Dams, *Dams and Development: A New Framework for Decision-Making* (London: Earthscan, 2000), p. 8.

14. Ibid.

15. Catherine M. Pringle, Mary C. Freeman, and Byron J. Freeman, "Regional Effects of Hydrologic Alterations on Riverine Macrobiota in the New World: Tropical-Temperate Comparisons," *BioScience*, September 2000, pp. 807–23.

16. Ibid.

17. David Dudgeon, "The Ecology of Tropical Asian Rivers and Streams in Relation to Biodiversity Conservation," *Annual Review of Ecological Systems*, vol. 31 (2000), pp. 239–63.

18. Ibid.

19. Ibid.

20. Ibid.

21. Ibid.

22. World Commission on Dams, op. cit. note 13, p. 9.

23. Margaret A. Palmer et al., "Linkages Between Aquatic Sediment Biota and Life above Sediments as Potential Drivers of Biodiversity and Ecological Processes," *BioScience*, December 2000, pp. 1062–75.

24. Ibid.

25. IUCN–The World Conservation Union, *Vision for Water and Nature: A World Strategy for Conservation and Sustainable Management of Water Resources in the 21st Century* (Gland, Switzerland, and Cambridge, U.K.: 2000).

26. J.M. King, R.E. Tharme, and M.S. de Villiers, eds., *Environmental Flow Assessments for Rivers: Manual for the Building Block Methodology* (Pretoria, South Africa: Water Research Commission, 2000); Republic of South Africa, "South African National Water Act No. 36 of 1998."

## TRANSBOUNDARY PARKS BECOME POPULAR (pages 108–09)

1. Arthur H. Westing, "International Conference on Transboundary Protected Areas as a Vehicle for International Cooperation," *Environmental Conservation*, vol. 25, no. 1 (1998).

2. Political obstacles delayed the creation of these twin parks until 1949; Paul Spencer Sochaczewski, "Across a Divide," *International Wildlife*, July/August 1999, p. 36.

3. Parks Canada, "About the Waterton/Glacier International Peace Park," at <parkscanada.pch.gc.ca/Parks/Alberta/Waterton_lakes/English/welcome 2_e.htm>, viewed 30 January 2002.

4. Sochaczewski, op. cit. note 2, pp. 34–41.

5. "Peace Parks Foundation Review," *Africa–Environment & Wildlife*, vol. 6, no. 1 (1998).

6. Dorothy Zbicz, "Global List of Complexes of Internationally Adjoining Protected Areas," Appendix 1, in Trevor Sandwith et al., *Transboundary Protected Areas for Peace and Co-operation*, Best Practice Protected Area Guidelines Series No. 7 (Gland, Switzerland, and Cambridge, U.K.: IUCN–The World Conservation Union and Cardiff University, 2001), pp. 55–75.

7. Ibid., p. 56. Protected areas are defined as government-designated areas that are at least 1,000 hectares in size.

8. Ibid.

9. Ibid.

10. Figure of 113 from ibid., p. 55; one third and 300 from Dorothy C. Zbicz, discussion with author, 30 January 2002.

11. Zbicz, op. cit. note 6, p. 56.

12. Ibid. Table 1 derived from the following sources: Jan Cerovsky, "Transfrontier Protected Areas Along the Former 'Iron Curtain' in Europe," in *Parks for Peace: Proceedings of the International Conference on Transboundary Protected Areas as a Vehicle for International Co-operation, Somerset West, South Africa, 16–18 September 1997* (Gland, Switzerland: IUCN, draft of 30 January 1998), pp. 117–20; World Wide Fund for Nature, "WWF Welcomes the Creation of the First Transboundary Park in the Balkans," press release (Gland, Switzerland: 2 February 2000); "Peace Parks: No Turning Back," *Africa Geographic*, December 2001/January 2002, pp. 86–87; Annette Lanjouw and José Kalpers, "Potential for the Creation of a Peace Park in the Virunga Volcano Region," in *Parks for Peace*, op. cit. this note, pp. 163–72; "Transboundary Biosphere Conservation, Development and Peace-Building: Lessons from the Altai Mountains," *Environmental Change and Security Project Report* (The Woodrow Wilson Center), summer 2001, p. 206; Deepak Gajurel, "Nepal and India Agree on Transborder Wildlife Conservation," *Environment News Service*, 3 May 1999; Arthur H. Westing, "A Transfrontier Reserve for Peace and Nature on the Korean Peninsula," in *Parks for Peace*, op. cit. this note, pp. 234–41; Juan J. Castro-Chamberlain, "Peace Parks in Central America: Successes & Failures in Implementing Management Cooperation," in ibid., pp. 49–60; Sochaczewski, op. cit. note 2, p. 38; Jocelyn Kaiser, "Bold Corridor Project Confronts Political Reality," *Science*, 21 September 2001, pp. 2196–99; Parks Canada, op. cit. note 3; Barbara Novovitch, "Feature: Peaceful Neighbors Eye U.S.-Mexico Peace Park," *Reuters*, 25 August 2000; Elizabeth Manning, "Russian Governor Favors Sea Park," *Anchorage Daily News*, 16 October 2001.

13. Zbicz, op. cit. note 10.

14. Ibid.

15. Ibid.; Westing, op. cit. note 1.

16. David McDowell, Director General, IUCN, "Opening Remarks," in *Parks for Peace*, op. cit. note 12, p. 25.

17. Ibid.

18. Lanjouw and Kalpers, op. cit. note 12.

19. Lawrence S. Hamilton, "Guidelines for Effective Transboundary Conservation: Philosophy and Best Practices," in *Parks for Peace*, op. cit. note 12, pp. 27–35.

20. Ibid.

21. "Peace Parks: Game Without Frontiers," *Zimbabwe Wildlife*, October-December 1998, p. 26; IUCN, "Transboundary Protected Areas Task Force (Parks for Peace)," at <wcpa.iucn.org/theme/parks/parks.html>, viewed 17 January 2002.

22. Gajurel, op. cit. note 12.

23. Sandwith et al., op. cit. note 6, p. 8.

24. Arthur H. Westing, "Establishment and Management of Transfrontier Reserves for Conflict Prevention and Confidence Building," *Environmental Conservation*, vol. 25, no. 2 (1998), pp. 91–94.

25. Ibid., p. 91.

26. IUCN, op. cit. note 21.

27. Westing, op. cit. note 24, p. 92.

28. Sandwith et al., op. cit. note 6, p. 9; Sochaczewski, op. cit. note 2, p. 38.

29. Sandwith et al., op. cit. note 6, pp. 19–21.

30. Peter Godwin, "Without Borders: Uniting Africa's Wildlife Reserves," *National Geographic*, September 2001, pp. 2–29.

31. "Declaration of Principles," in *Parks for Peace*, op. cit. note 12, p. 15.

32. Sandwith et al., op. cit. note 6, p. 14.

# Notes

33. Ibid.
34. "Peace Parks: Game Without Frontiers," op. cit. note 21.

## SEMICONDUCTORS HAVE HIDDEN COSTS (pages 110–11)

1. Semiconductor Industry Association (SIA), "Industry Facts and Figures," at <www.semi chips.org/ind_facts.cfm>, viewed 6 January 2002.
2. Ibid.
3. Ibid.
4. SIA, "World Market Sales and Shares for 1982–1990," and "World Market Sales and Shares for 1991–2000," at <www.semichips.org/pre_statistics.cfm>, viewed 18 September 2001.
5. SIA, "Semiconductor Forecast Summary 2001-4," November 2001, at <www.semichips.org/down loads/Forecast Summary - Fall '01.pdf>, viewed 2 January 2002.
6. G. Dan Hutcheson and Jerry D. Hutcheson, "Technology and Economics in the Semiconductor Industry," *Scientific American,* October 1997.
7. Semiconductor International Capacity Statistics, "Statistics Report—3rd Quarter 2001: Integrated Circuit Wafer-Fab Capacity," and "Statistics Report—3rd Quarter 2001: Integrated Circuit Wafer-Fab Utilisation," at <www.semichips.org/pre_statistics.cfm>, viewed 2 January 2002.
8. Joseph LaDou and Timothy Rohm, "The International Electronics Industry," *International Journal of Occupational and Environmental Medicine,* vol. 4, no. 1 (1998), p. 1.
9. Bill Richards, "Semiconductor Plants Aren't Safe and Clean as Billed, Some Say," *Wall Street Journal,* 5 October 1998.
10. Environmental Defense, at <www.scorecard.org/env-releases/land/rank-counties.tcl>, viewed 8 January 2002.
11. SIA, op. cit. note 1; LaDou and Rohm, op. cit. note 8, p. 1.
12. M. B. Schenker et al., "Association of Spontaneous Abortion and Other Reproductive Effects with Work in the Semiconductor Industry," *American Journal of Industrial Medicine,* December 1995, pp. 639–59.
13. Ron Chepesiuk, "Where the Chips Fall: Environmental Health in the Semiconductor Industry," *Environmental Health Perspectives,* vol. 107, no. 9 (1999), pp. A452–57.
14. Ibid.
15. P. J. Huffstutter, "IBM Settles Suit Alleging Plants Caused Birth Defects," *Los Angeles Times,* 24 January 2001.
16. Craig Wolf, "IBM Settles Chemical Suit," *Poughkeepsie Journal,* 23 January 2001.
17. Commission of the European Communities, *Proposal for a Directive of the European Parliament and of the Council on Waste Electrical and Electronic Equipment* (Brussels: 12 June 2000), p. 4.
18. Ibid.
19. Silicon Valley Toxics Coalition (SVTC), *Poison PCs and Toxic TVs: California's Biggest Environmental Crisis That You've Never Heard Of* (San Jose, CA: June 2001), p. 2.
20. National Safety Council, *Electronic Product Recovery and Recycling Baseline Report: Recycling of Selected Electronic Products in the United States* (Washington, DC: 1999).
21. Ibid.
22. Lou Hirsh, "The Next Environmental Crisis: Techno-Trash," *NewsFactor Network,* 29 May 2001, at <www.newsfactor.com/perl/printer/10034>, viewed 3 January 2002.
23. "Making Japan a Recycling Society: New Appliance Recycling Law to Take Effect," *Trends in Japan,* Japan Information Network, 23 March 2001, at <jin.jcic.or.jp/trends00/honbun/tj 010323.html>, viewed 6 January 2002.
24. European Environmental Bureau, "Press Release: Council Fails to Follow Parliament on WEEE Directive," 8 June 2001, at <www.eeb.org/press/press_release_WEEE_08_06.htm>, viewed 3 January 2002.
25. Commission of the European Communities, op. cit. note 17, p. 15–17.
26. SVTC, op. cit. note 19, p. 9.
27. Ibid., p. 3.
28. Ibid., p. 9. Toxicity data from Environmental Defense, at <www.scorecard. org>, viewed 9 January 2002.
29. SVTC, op. cit. note 19, p. 9. Toxicity data from Lawrence Tierney, Jr., Stephen McPhee, and Maxine A. Papadakis, eds., *Current Medical Diagnosis and Treatment 2002,* 41st ed. (New York: McGraw-Hill, 2002), p. 1632.

## TOXIC WASTE LARGELY UNSEEN
(pages 112–13)

1. Jonathan Krueger, "What's to Become of Trade in Hazardous Wastes? The Basel Convention One Decade Later," *Environment*, November 1999, p. 13.
2. U.S. Bureau of the Census, *International Data Base*, electronic database, Suitland, MD, updated 10 May 2000; no end in sight from Jonathan Krueger, Chemicals and Waste Management Programme, U.N. Institute for Training and Research, Geneva, discussion with author, 4 January 2002.
3. U.N. Environment Programme (UNEP), "Basel Convention on the Control of Transboundary Movements of Hazardous Wastes and Their Disposal Adopted by the Conference of the Plenipotentiaries on 22 March 1989, Entry into Force 5 May 1992," at <www.unep.ch/basel/text/con-e.pdf> viewed 4 January 2002; idem, "Basic Description of the Basel Convention," general information leaflet, at <www.unep.ch/basel/about.html#basic>, viewed 4 January 2002.
4. U.N. Secretary-General, "Message to the Fifth Meeting of the Parties to the Basel Convention" (Basel: 6 December 1999).
5. Krueger, op. cit. note 1, p. 13; Commission on Human Rights (CHR), U.N. Economic and Social Council, *Adverse Effects of the Illicit Movement and Dumping of Toxic and Dangerous Products and Wastes on the Enjoyment of Human Rights*, Report Submitted by the Special Rapporteur on Toxic Waste, Mrs. Fatma-Zohra Ouhachi-Vesely (Geneva: 19 January 2001), p. 26.
6. Gilbert T. Bergquist et al., *Chemical and Pesticides Results Measures* (Tallahassee, FL: U.S. Environmental Protection Agency (EPA) and Institute of Science and Public Affairs, Florida State University, February 2001), p. 69; Krueger, op. cit. note 1, p. 13.
7. "Table 1: Total Amount of Hazardous Wastes and Other Wastes Generated in 1998 (as reported by Parties)," in Secretariat of the Basel Convention, *Compilation Part II: Reporting and Transmission of Information Under the Basel Convention for the Year 1998 (Statistics on Generation and Transboundary Movements of Hazardous Wastes and Other Wastes)*, Basel Convention Series/SBC No. 00/05 (Geneva: December 2000). Note: other wastes include household wastes and residues from the incineration of household wastes, as specified in Annex II of the Basel Convention. Annex II wastes require special consideration since they can contain small quantities of hazardous waste.
8. Secretariat of the Basel Convention, "Country Fact Sheets (1999)," at <www.unep.ch/basel/pub/99cfs.pdf>, viewed 4 January 2002.
9. Data for 1998 from "Table 1," op. cit. note 7; 1999 data from Secretariat of the Basel Convention, op. cit. note 8, pp. 64, 370.
10. "Introduction," in Secretariat of the Basel Convention, op. cit. note 7, pp. 1–4.
11. UNEP, "Basic Description," op. cit. note 3.
12. Jennifer Clapp, *Toxic Exports: The Transfer of Hazardous Wastes from Rich to Poor Countries* (Ithaca, NY: Cornell University Press, 2001), pp. 45, 67–68.
13. Secretariat of the Basel Convention, "Ratifications of the Ban Amendment (updated 2 January 2002)," at <www.unep.ch/basel/ratif/ratif.html#banratif>, viewed 4 January 2002.
14. Pat Phibbs, "Ratification of Basel Accord Without Ban On Exports Opposed by Environment Coalition," *International Environment Reporter*, 15 August 2001, p. 687.
15. Secretariat of the Basel Convention, "Basel Convention on the Control of Transboundary Movements of Hazardous Wastes and Their Disposal: Status of Ratification/Accession/Acceptance/Approval as of 2 January 2002," at <www.unep.ch/basel/ratif/ratif.html#conratif>, viewed 4 January 2002.
16. CHR, op. cit. note 5, p. 27.
17. Ibid.
18. Clapp, op. cit. note 12, p. 152.
19. Krueger, op. cit. note 1, p. 13.
20. CHR, op. cit. note 5, p. 26.
21. Texas Center for Policy Studies (TCPS), *The Generation and Management of Hazardous Wastes and Transboundary Hazardous Waste Shipments between Mexico, Canada and the United States, 1990–2000* (Austin, TX: May 2001); Martin Mittelstaedt, "Canada Becomes Haven for Toxic Waste," (Toronto) *Globe and Mail*, 18 June 2001.
22. TCPS, op. cit. note 21, p. 52.
23. Mittelstaedt, op. cit. note 21.
24. Jim Puckett, "The Basel Treaty's Ban on Hazardous Waste Exports: An Unfinished Success Story," *International Environment Reporter*, 6 December 2000, p. 984.

25. UNEP, "Illegal Traffic in Hazardous Wastes Under the Basel Convention," general information leaflet, at <www.unep.ch/basel/pub/pub.html>, viewed 4 January 2002; Jonathan Krueger, *International Trade and the Basel Convention* (London: Earthscan, 1999), pp. 87–95.

26. Puckett, op. cit. note 24.

27. CHR, op. cit. note 5, p. 3.

28. "Table 2: Transboundary Movement of Hazardous Wastes and Other Wastes for Disposal among all Reporting Parties in 1998," and "Table 3: Transboundary Movement of Hazardous Wastes and Other Wastes for Recycling among all Reporting Parties in 1998," in Secretariat of the Basel Convention, op. cit. note 7.

29. "Addendum," in CHR, op. cit. note 5, pp. 3–4.

30. Clapp, op. cit. note 12, pp. 104–25; CHR, op. cit. note 5, pp. 3, 15.

31. Greenpeace International, "Greenpeace Calls on Shipping Industry to Stop Using Asia as Dumping Ground," press release (Amsterdam: 15 February 2001).

32. Henry J. Holcombe, "Shippers Are Urged to Cease Dangerous Recycling Methods," *Philadelphia Inquirer*, 11 September 2001.

33. Martine Vrijheid, "Health Effects of Residence Near Hazardous Waste Landfill Sites: A Review of Epidemiologic Literature," *Environmental Health Perspectives*, March 2000, pp. 101–12.

34. Susanna Jacona Salafia, "Mothers Near Landfills Risk Malformed Babies," *Environmental News Service*, 3 November 1998.

35. Michael Berry and Frank Bove, "Birth Weight Reduction Associated with Residence Near a Hazardous Waste Landfill," *Environmental Health Perspectives*, August 1997, pp. 105–08.

36. Estimate of hazardous waste generation per day derived from Krueger, op. cit. note 1, p. 13.

## RIO TREATIES POST SOME SUCCESS
(pages 114–15)

1. Table 1 from the following: Convention on Biological Diversity, at <www.biodiv.org>; *CBD News*, January/March 2001; U.N. Framework Convention on Climate Change (UNFCCC), at <www.unfccc.de>; Jon Hanks et al., *Earth Negotiation Bulletin*, 30 July 2001, at <www.iisd.ca/climate/cop6bis>; Sharon Taylor, Librarian, UNFCCC Secretariat, e-mail to author, 25 October 2001; Convention to Combat Desertification, at <www.unccd.int>; *Down to Earth: Newsletter of the Convention to Combat Desertification*, December 2000, pp. 1–3; "Desertification: Parties to U.N. Convention to Open Meeting in Geneva," *U.N. Wire*, 1 October 2001; Convention on Straddling Fish Stocks, at <www.un.org/Depts/los/convention_agreements/convention_overview_fish_stocks.htm>; "Straddling Stocks Agreement Important to Large Migratory Fish," *Dispatches*, February 2000; Prior Informed Consent (PIC) Convention, at <www.pic.int>; Persistent Organic Pollutants (POPs) Treaty, at <www.chem.unep.ch/sc/>; WWF's Global Toxic Chemicals Initiative, "Summary of Key Elements in the Global POPs Treaty," 14 December 2000, at <www.worldwildlife.org/toxics/progareas/pop/treaty_summary.pdf>; U.N. Environment Programme, "Stockholm Convention on POPs," *UNEP Chemicals*, June 2001, at <www.chem.unep.ch>.

2. Rosalie Gardiner and Zoe Hatherly, "Conventions Briefing Paper," UNED Forum, at <www.earthsummit2002.org/es/issues/Conventions/rio_conventions.htm>, viewed on 24 January 2002.

3. David G. Victor, Kal Raustiala, and Eugene B. Skolnikoff, eds., *The Implementation and Effectiveness of International Environmental Commitments: Theory and Practice* (Cambridge, MA: The MIT Press, 1998), p. 660–61; Andrew C. Revkin, "Global Warming Impasse is Broken," *New York Times*, 11 November 2001.

4. Victor, Raustiala, and Skolnikoff, op. cit. note 3, p. 683.

5. Eric Pianin, "Warming Pact a Win for European Leaders," *Washington Post*, 11 November 2001; David Buchanan, "UN Meeting Reaches Accord on Global Warming," *Financial Times*, 12 November 2001.

6. Andrew C. Revkin, "Negotiators Push to Settle World Treaty on Warming," *New York Times*, 10 November 2001; Pianin, op. cit. note 5.

7. Reuters, "Global-Warming Talks Gain With Accord on Compliance," *New York Times*, 8 November 2001; Buchanan, op. cit. note 5.

8. Steven R. Ratner, "International Law: The Trials of Global Norms," *Foreign Policy*, spring 1998, p. 67; Victor, Raustiala, and Skolnikoff, op. cit. note 3, p. 680. Other treaties that include trade provisions are the Basel Convention on the Transboundary Movement of Hazardous Waste, the Convention

on International Trade in Endangered Species, PIC, and POPs.

9. Union of International Organizations, *Yearbook of International Organizations* (Munich: K. G. Sauer Verlag, 2000/2001), Appendix 3, Table 1.

10. Aarhus Convention, at <www.unece.org/env/pp/ctreaty.htm>.

11. Peter Veit, World Resources Institute, discussion with author, 1 November 2001.

12. Scott Barrett, "If Not Kyoto, What?" *SAISPHERE*, 2001, p. 30; Pianin, op. cit. note 5.

13. Edith Brown Weiss and Harold K. Jacobson, "Why Do States Comply with International Agreements?" *Human Dimensions Quarterly*, spring 1996, pp. 1–5.

14. Ashley Mattoon, "Evidence of Global Extinction Crisis Builds," *World Watch*, January/February 2001, pp. 8–9; Jim Motavalli and Jennifer Bogo, "The Last of Their Kind," *E: The Environmental Magazine*, May 1999, pp. 28, 29, 35; Hilary French, *Vanishing Borders* (New York: W.W. Norton & Company, 2000), pp. 89, 145.

15. Carsten Helm and Detlef Sprinz, "Measuring the Effectiveness of International Environmental Regimes," *Journal of Conflict Management,* October 2000, pp. 630–52.

## FOREIGN AID SPENDING FALLS
(pages 118–19)

1. See especially Chapters 2 and 33 of United Nations, *Agenda 21: the United Nations Programme of Action from Rio* (New York: U.N. Department of Public Information, undated).

2. United Nations, op. cit. note 1.

3. Official development assistance in 1992 of $60.42 billion (in current dollars) from Organisation for Economic Co-operation and Development (OECD), Development Assistance Committee, *Development Co-operation 1993* (Paris: 1994), pp. 168–69.

4. United Nations, op. cit. note 1, Chapter 33.

5. "Media Kit: Addressing the Quantity and Quality of Development Aid," International Conference on Financing for Development, Monterrey, Mexico, 18–22 March 2002.

6. Global Environment Facility (GEF), *GEF Contributions to Agenda 21: The First Decade* (Washington, DC: June 2000), pp. 6–9; idem, *Joint Summary of the Chairs*, GEF Council Meeting, 9–11 May 2001 (Washington, DC: 15 May 2001), p. 2.

7. GEF, *GEF Contributions*, op. cit. note 6, pp. 3–5.

8. GEF, "GEF Projects—Allocations and Disbursements," 3 October 2001, paper submitted to Meeting on the Third Replenishment of the GEF Trust Fund, 11–12 October 2001.

9. Treaty secretariat budgets from U.N. Environment Programme (UNEP), "International Environmental Governance: Multilateral Environmental Agreements (MEAs)," paper prepared for the Open-Ended Intergovernmental Group of Ministers or their Representatives on International Environmental Governance, Montreal, Canada, 30 November–1 December 2001, pp. 41–43, and from Hilary French and Lisa Mastny, "Controlling International Environmental Crime," in Lester R. Brown et al., *State of the World 2001* (New York: W.W. Norton & Company, 2001), p. 171. UNEP's spending for the two-year period 2000–01 was $210.54 million, according to "Status of the Environment Fund and other Sources of Funding for the United Nations Environment Programme," Governing Council of UNEP, Seventh Special Session, Cartagena, Colombia, February 2002, p. 2. This figure includes the resources from the U.N. regular budget, from the Environment Fund, and from Trust Funds, along with counterpart contributions; $183 million of this total was devoted to program resources and the remainder to management and administrative support.

10. The Environmental Protection Agency received just over $7.8 billion in fiscal years (FY) 2000 and 2001 and is expected to receive a similar amount in 2002, according to "FY2002 Annual Performance Plan and Congressional Justification Appropriation (EPA's Proposed Budget)," at <www.epa.gov/ocfo/budget/2002/2000cj.htm>, viewed 12 October 2001. The U.S. military budget amounted to $300,767 million in FY2000 and to $311,271 million in FY2001, according to <www.whitehouse.gov/omb/budget/fy2002/budget.html>, viewed 12 October 2001. World military expenditures of $784 billion in 2000 from Stockholm International Peace Research Institute, "World and Regional Military Expenditure Estimates, 1991–2000," at <projects.sipri.se/milex/mex_wnr_table.html>, viewed 11 October 2001.

11. Sheryl Gay Stolberg, "AIDS Fund Falls Short of Goal and U.S. Is Given Some Blame," *New York Times*, 13 February 2002.

12. Ibid.

13. U.N. Millennium Declaration, at <www.un.org/millennium/declaration/ares552e.htm,> viewed 24 January 2002.

14. Ernesto Zedillo et al., "Recommendations of the High-level Panel on Financing for Development," commissioned by the Secretary-General of the United Nations, New York, 22 June 2001, p. 34.

15. Ibid., p. 19.

16. Joseph Kahn, "Britain Urges U.S. to Expand Worldwide Antipoverty Programs," *New York Times*, 18 December 2001.

## CHARITABLE GIVING WIDESPREAD
(pages 120–21)

1. Table 1 based on the following: Canada from Michael Hall et al., *Caring Canadians, Involved Canadians: Highlights from the 2000 National Survey of Giving, Volunteering and Participating* (Ottawa: Statistics Canada, August 2001), p. 10. U.S. estimate is for 2000 from Independent Sector, "Giving and Volunteering in the United States 2001, Key Findings," at <www.Independent Sector.org>, which reports information for households. This statistic has been divided by 1.7 (the average number of adults per American household according to Bureau of Census, "American Housing Survey-1999," at <www.census.gov/hhes/www/housing/ahs/99dtchrt/tab2-9.html>) for easier comparison with per person statistics from other nations. The share of the population that gives, however, indicates the share of households that gives. Netherlands from Rene Bekkers and Theo Schuyt, "Giving in the Netherlands 2001: Summary of Practical Findings of the Biannual Research Report on Philanthropy and Volunteering in the Netherlands," Center for the Study of Philanthropy and Volunteering, Amsterdam, August 2001, at <filantropie.scw.vu.nl/gin/english.htm>, p. 3. United Kingdom from "Charitable Giving—The Tide Has Turned. The Fall and Rise of Charitable Giving 1995–2000," *Research Quarterly* (National Council for Voluntary Organizations, London), June 2001. Japan from "Giving and Tax Deductions in Japan," at <www.igc.org/ohdakefoundation/npo/npotax.htm>, and from Economic Planning Agency, Government of Japan, "White Paper on the National Lifestyle," at <www5.cao.go.jp/2000/c/1110wp-seikatsu-e/main.html>, November 2000. This source reports information for households, and this statistic has been divided by 1.7 (assuming the average number of adults per Japanese household is roughly the same as the U.S. figure) for comparison purposes. France from Jacques Malet, *La Générosité des Français: Etude sur les Dons Déclarés* (Paris: Fondation de France, 2001), p. 18.

2. Kathy Steinberg, Assistant Director of Research, The Center on Philanthropy at Indiana University, discussion with author, 7 January 2002.

3. Bekkers and Schuyt, op. cit. note 1, p. 1.

4. Steinberg, op. cit. note 2.

5. Hall et al., op. cit. note 1; Steinberg, op. cit. note 2.

6. AAFRC Trust for Philanthropy, "Giving USA 2001," powerpoint presentation, sent to author by Joan Ward, 29 November 2001; Steinberg, op. cit. note 2.

7. AAFRC, op. cit. note 6; Hall et al., op. cit. note 1, p. 10.

8. Hall et al., op. cit. note 1, p. 10.

9. AAFRC, op. cit. note 6.

10. Ibid.

11. Ibid.

12. Hall et al., op. cit. note 1, p. 12.

13. "Giving and Tax Deductions in Japan," op. cit. note 1.

14. Ibid.; Economic Planning Agency, op. cit. note 1.

15. AAFRC, op. cit. note 6.

16. Hall et al., op. cit. note 1, p. 11.

17. Ibid., pp. 18–19.

18. Ibid., pp. 17, 20; tendency found in Europe from Bekkers and Schuyt, op. cit. note 1, p. 4; Organisation for Economic Co-operation and Development (OECD), *The Well-being of Nations: The Role of Human and Social Capital* (Paris: 2001), p. 34.

19. Hall et al., op. cit. note 1, p. 20; Independent Sector, op. cit. note 1.

20. OECD, op. cit. note 18, p. 34.

21. "Charitable Giving—The Tide Has Turned," op. cit. note 1; Bekkers and Schuyt, op. cit. note 1.

22. Hall et al., op. cit. note 1, p. 22.

23. AAFRC, op. cit. note 6.

24. Tendency found in Europe from Bekkers and Schuyt, op. cit. note 1, p. 2.

25. "Charitable Giving—The Tide Has Turned," op. cit. note 1.

26. United Kingdom from ibid.; United States from AAFRC, op. cit. note 6.

27. Hall et al., op. cit. note 1, p. 26.

28. John O'Neil, "Charities Get a Big Helping of

Uncertainty," *New York Times*, 12 November 2001; Independent Sector, "Charitable Giving: September 11th and Beyond," press release (Washington, DC: 23 October 2001).

29. Hall et al., op. cit. note 1, p. 11.

30. Independent Sector, op. cit. note 1.

31. Hall et al., op. cit. note 1, p. 28; Independent Sector, op. cit. note 1.

32. Hall et al., op. cit. note 1, p. 28.

33. Ibid.

### CRUISE INDUSTRY BUOYANT (pages 122–23)

1. John Dearing, G.P. Wild (International) Limited, Haywards Heath, West Sussex, U.K., e-mail to author, 7 December 2001.

2. Worldwatch estimates derived from data provided by Dearing, op. cit. note 1, and by Rosa Songel, Economic and Statistical Measurement of Tourism, World Tourism Organization (WTO), e-mail to author, 17 April 2001.

3. G.P. Wild (International) Limited, *Implications of Fleet Changes for Cruise Market Prospects to 2010* (Haywards Heath, West Sussex, U.K.: August 2001).

4. Cruise Lines International Association (CLIA), *Cruise Industry Source Book* (New York: 2001).

5. WTO, *Tourism Market Trends: World Overview and Tourism Topics* (Madrid: May 2001), p. 68.

6. Floating cities from "Cruise Ship Dumping Sparks Interest," *Associated Press*, 3 December 1999; passengers and crew from International Transport Workers' Federation (ITF), "Cruise Ship Campaign," at <www.itf.org.uk/seafarers/cruise_ships/index.htm>, viewed 5 January 2002; amenities from WTO, op. cit. note 5, p. 68.

7. CLIA, *Cruise Industry Overview* (New York: September 2001); G.P. Wild (International) Limited, ""Supply Side Activities," *International Cruise Market Monitor*, vol. 7, no. 3 (2001), p. 9.

8. Estimate of 76 percent from ITF, op. cit. note 6; revenues from corporate Web sites, at <www.carnival.com>, <www.poprincess.com>, <www.royalcaribbean.com>, and <www.starcruises.com>, viewed 19 December 2001.

9. P&O Princess plc, "P&O Princess Cruises plc and Royal Caribbean Cruises Ltd. Combine to Create the World's Largest Cruise Vacation Group," press release (London: 20 November 2001).

10. Dearing, op. cit. note 1; Asia and the Pacific from WTO, op. cit. note 5, pp. 61–62.

11. CLIA, op. cit. note 7.

12. "Alaska Governor Proposes New Cruise-Ship Controls," *Reuters*, 13 March 2001; Gregg Fields, "Population Pollution is a Concern for Cruise Company," *Environmental News Network*, 19 July 2000.

13. Business Research and Economic Advisors (BREA), *The Contribution of the North American Cruise Industry to the U.S. Economy in 2000* (Exton, PA: October 2001), p. 3.

14. International Council of Cruise Lines (ICCL), *Cruise Industry FAQs* (Arlington, VA: 2001); low fees from Mary B. Uebersax, "Indecent Proposal: Cruise Ship Pollution in the Caribbean," August 1996, at <www.planeta.com/96/0896cruise.html>, viewed 26 October 2001.

15. Uebersax, op. cit. note 14; Robert Wood, "Caribbean Cruise Tourism: Globalization at Sea," *Annals of Tourism Research*, vol. 27, no. 2 (2000), pp. 349–50; ICCL, op. cit. note 14.

16. ICCL, op. cit. note 14.

17. Douglas Frantz, "Gaps in Sea Laws Shield Pollution by Cruise Lines," *New York Times*, 3 January 1999.

18. Carnival from Nanette Byrnes, "Few Icebergs in the Horizon," *Business Week*, 14 June 1998, p. 83; Royal Caribbean from Frantz, op. cit. note 17.

19. Wood, op. cit. note 15, p. 351.

20. "The Tourism Industry: A Report for the World Summit on Sustainable Development, Johannesburg, September 2002," compiled by World Travel & Tourism Council with International Federation of Tour Operators, International Hotel & Restaurant Association, and International Council of Cruise Lines (draft), December 2001, p. 37; nationalities from ITF, op. cit. note 6.

21. Christopher Reynolds and Dan Weikel, "For Cruise Ship Workers, Voyages Are No Vacations," *Los Angeles Times*, 6 June 2000.

22. Frantz, op. cit. note 17.

23. Kira Schmidt, *Cruising for Trouble: Stemming the Tide of Cruise Ship Pollution* (San Francisco: Bluewater Network, March 2000).

24. Yareth Rosen, "Cruise Ships Cited for Excessive Smoke in Alaska," *Reuters*, 23 August 2000.

25. Discharge total from Martha Honey, *Ecotourism and Sustainable Development: Who Owns Paradise?* (Washington, DC: Island Press, 1999), p. 40.

26. See International Maritime Organization conven-

tions, available at <www.imo.org>.

27. Frantz, op. cit. note 17; "Royal Caribbean Sentenced for Ocean Dumping," *Reuters*, 4 November 1999.

28. Alaska Department of Environmental Conservation, *Alaska Cruise Ship Initiative: Part 2 Report (June 1, 2000 to July 1, 2001)* (Juneau: 2001), pp. 13–14.

29. S. H. Smith, "Cruise Ships: A Serious Threat to Coral Reefs and Associated Organisms," *Ocean and Shoreline Management*, vol. 11 (1988), pp. 231–48; Cayman Islands from Polly Pattullo, *Last Resorts: The Cost of Tourism in the Caribbean* (London: Cassell, 1996), p. 110.

30. Recycling from Uebersax, op. cit. note 14.

31. "Cruise Line Cleans Up Its Wake," *Environmental News Network*, 6 August 2001; Uebersax, op. cit. note 14.

32. ICCL, "ICCL Industry Standard E-01-01: Cruise Industry Waste Management Practices and Procedures," adopted 11 June 2001, at <www.iccl.org/policies/environmentalstandards.pdf>.

33. Ibid.

34. U.N. Environment Programme, *Environmental Codes of Conduct for Tourism* (Nairobi: 1995).

35. International Association of Antarctic Tour Operators, at <www.iaato.org >, viewed 10 October 2001; "Antarctic Tourism Tests Fragile Ecosystem," *Reuters*, 23 February 1999.

36. Rainforest Alliance, "About SmartVoyager," at <www.rainforest-alliance.org/programs/sv/index.html>, viewed 21 August 2001; Jorge Peraza-Breedy, Sustainable Tourism Program, Rainforest Alliance, San José, Costa Rica, e-mail to author, 21 August 2001.

37. "Alaska to Regulate Pollution from Cruise Ships," *Reuters*, 11 June 2001.

38. Yereth Rosen, "Alaska to Cruise Ships: We're Not Your Sewer," *Christian Science Monitor*, 12 July 2001; "Alaska to Regulate Pollution from Cruise Ships," op. cit. note 37; "Alaska Hostile to Cruise Line's Apology," *Associated Press*, 27 August 1999.

## ECOLABELING GAINS GROUND
(pages 124–25)

1. D. J. Caldwell, *Ecolabeling and the Regulatory Framework: A Survey of Domestic and International Fora,* report prepared for the Consumer's Choice Council, Washington, DC, 30 October 1998; Robert M. Abbott, *Eco-labeling and the Green Economy: Strategic Options for British Columbia* (Vancouver, BC, Canada: Abbot Strategies, 24 November 2000), p. 5.

2. Consumers Union, "Eco-labels," at <www.ecolabels.org>, viewed 25 January 2002.

3. Caldwell, op. cit. note 1.

4. Forest Stewardship Council (FSC), at <www.fscoax.org>, viewed 17 January 2002.

5. Abbott, op. cit. note 1, p. 6.

6. Table 1 based on the following: Abbott, op. cit. note 1, p. 16; FSC, op. cit. note 4; idem, "Forests Certified by FSC-Accredited Certification Bodies," 13 February 2002, at <www.fscoax.org>, viewed 3 March 2002; Rainforest Alliance, "Conservation Agriculture Network," at <www.rainforestalliance.org/programs/cap/faq.html>, viewed 17 January 2002; Marine Stewardship Council, at <www.msc.org>, viewed 17 January 2002; Smithsonian Migratory Bird Center, at <natzoo.si.edu/smbc/Research/Coffee/coffee.htm>, viewed 4 February 2002; Green-e, at <www.green-e.org>, viewed 4 February 2002; Blue Flag from <www.blueflag.org> and from Graham Ashworth, "The Blue Flag Campaign," *Naturopa*, no. 88 (1998), p. 21.

7. National Wildlife Federation (NWF), *Guarding the Green Choice: Environmental Labeling and the Rights of Green Consumers* (Washington, DC: 1996), p. 6; Blue Angel, at <www.blauer-engel.de>, viewed 4 February 2002.

8. NWF, op. cit. note 7.

9. Abbott, op. cit. note 1, p. 5.

10. Ecolabelling Norway, at <www.ecolabel.no>, viewed 4 February 2002; European Union Ecolabel home page, at <europa.eu.int/comm/environment/ecolabel/index.htm>, viewed 4 February 2002.

11. European Union Ecolabel, "Frequently Asked Questions," at <europa.eu.int/comm/environment/ecolabel/helpfaq.htm>, viewed 4 February 2002.

12. Ibid.

13. Jim Motavelli, "The Color of Money," *E Magazine*, July/August 2000.

14. Consumers Union, op. cit. note 2.

15. Ibid.

16. Erin E. Dooley, "The Consumers Union Guide to Environmental Labels," *Environmental Health Perspectives*, September 2001, p. A-419.

17. Consumers Union, op. cit. note 2.

18. Synergy, "Tourism Certification: An Analysis of Green Globe 21 and Other Tourism Certification Programs," report prepared for WWF International (Gland, Switzerland: August 2000).

19. Rainforest Alliance, "Sustainable Tourism Stewardship Council," at <www.rainforest-alliance.org/programs/sv/stsc.html>, viewed 4 February 2002.

20. "Don't be Fooled: The Limits of Eco-labels," *Energy Ideas* (Government Purchasing Project), winter 1996.

21. "Programs That Help Consumers Identify 'Energy-Efficient' Products," *Energy Ideas* (Government Purchasing Project), winter 1996.

22. "Don't be Fooled," op. cit. note 20.

23. Ibid.

24. Dooley, op. cit. note 16.

25. Synergy, op. cit. note 18.

26. Caldwell, op. cit. note 1; Dooley, op. cit. note 16; Consumers Union, op. cit. note 2.

27. Dooley, op. cit. note 16; Consumers Union, op. cit. note 2.

28. "Don't be Fooled," op. cit. note 20.

29. Ibid.

30. Ibid.

## PESTICIDE SALES REMAIN STRONG
(pages 126–27)

1. Rob Bryant, Agranova, e-mail to author, 17 July 2001, with data adjusted for inflation using U.S. Commerce Department, Bureau of Economic Analysis, *U.S. Implicit GNP Price Deflator*, at <www.bea.doc.gov/bea/dn/st-tabs.htm>.

2. Ibid.
3. Ibid.
4. Ibid.
5. Ibid.
6. Ibid.

7. "Non-Ag Pesticides Market Growing," *Global Pesticide Campaigner*, August 2000, p. 7.

8. Ibid.
9. Ibid.
10. Ibid.
11. Bryant, op. cit. note 1.
12. Ibid.
13. Ibid.
14. Ibid.
15. Ibid.
16. Ibid.
17. Ibid.

18. Ibid.
19. Ibid.

20. These estimates are of volume of active ingredient; 2.6 million tons is an estimate for 1997 from Arnold L. Aspelin and Arthur H. Grube, *Pesticide Industry Sales and Usage: 1996 and 1997 Market Estimates* (Washington, DC: U.S. Environmental Protection Agency, 1999), p. 10; 3.5 million tons and 10-fold increase since 1950 is from David Pimentel, Cornell University, discussion with author, 12 December 2001.

21. ETC Group, "Globalization, Inc., Concentration in Corporate Power: The Unmentioned Agenda," *Communique* (Winnipeg, MN, Canada), July/August 2001, p. 9.

22. Ibid.
23. Ibid.

24. U.N. Food and Agriculture Organization (FAO), *FAOSTAT Statistics Database*, at <apps.fao.org>, updated 4 December 2001; data adjusted for inflation using U.S. Commerce Department, op. cit. note 1.

25. Carl Smith, "Pesticide Exports from U.S. Ports, 1997–2000," *International Journal of Occupational and Environmental Health*, October/December 2001.

26. World Health Organization in collaboration with U.N. Environment Programme, *Public Health Impact of Pesticides Used in Agriculture* (Geneva: 1990).

27. Ibid.

28. FAO, "FAO Warns: Toxic Pesticide Waste Stocks Dramatically Higher than Previously Estimated—Calls on Countries and Industry to Speed Up Disposal," press release (Rome: 9 May 2001).

29. Ibid.

30. J. N. Pretty et al., "An Assessment of the Total External Costs of UK Agriculture," *Agricultural Systems*, August 2000, pp. 113–36; Jules Pretty, University of Essex, Colchester, U.K., e-mail to author, 26 September 2001.

31. Montague Yudelman et al., *Pest Management and Food Production: Looking to the Future* (Washington, DC: International Food Policy Research Institute, September 1998), pp. 13–16.

32. Ibid., p. 13.

33. David Pimentel, "Protecting Crops," in Walter C. Olson, ed., *The Literature of Crop Science* (Ithaca, NY: Cornell University Press, 1995); one fifth from Bryant, op. cit. note 1.

34. Jules Pretty and Rachel Hine, *Reducing Food Poverty with Sustainable Ag: A Summary of New Evidence* (Colchester, U.K.: SAFE–World Research Project, University of Essex, February 2001), p. 131.

35. Jules Pretty, University of Essex, *AgriCulture: Communities Shaping the Land and Nature*, unpublished manuscript sent to author, 20 September 2001.

36. "Canadian IPM Program Cuts Pesticide Use by 40 Percent," *Gempler's IPM Solutions*, electronic newsletter, at <www.ipmalmanac.com/solutions>, December 2001.

## BIOTECH INDUSTRY GROWING
(pages 130–31)

1. Global estimate of revenues based on Ernst & Young, *Focus on Fundamentals: The Biotechnology Report* (New York: 2001), p. 9, on Ernst & Young, *European Life Sciences 2001: Integration* (New York: 2001), p. 8, on "India's Fermentation Queen," *The Economist*, 1 September 2001, p. 58, and on estimate for Australia from Charlie Craig, Ernst & Young, New York, discussion with author, 16 January 2002.

2. Table 1 based on the following: pharmaceuticals from Ernst & Young, *Focus on Fundamentals*, op. cit. note 1, pp. 12, 73, and from The Catalyst Group, "The Biotechnology Revolution: Biocatalysis Toolbox Design for Competitive Advantage," press release (Spring House, PA: 15 March 2001); agriculture from Clive James, "Global Review of Commercialized Transgenic Crops: 2001," *ISAAA Briefs No. 24: Preview*, January 2002, from Sheryl Gay Stolberg, "Breakthrough in Pig Cloning Could Aid Organ Transplants," *New York Times*, 4 January 2002, and from Carol Kaesuk Yoon, "If It Walks and Moos Like a Cow, It's a Pharmaceutical Factory," *New York Times*, 1 May 2000; information technologies from Karen J. Watkins, "Bioinformatics," *Chemical & Engineering News*, 19 February 2001, pp. 29–45, and from Yaakov Benenson, "Programmable and Autonomous Computing Machine Made of Biomolecules," *Nature*, 22 November 2001, pp. 430–34; human life applications from Nicholas Wade, "In Tiny Cells, Glimpses of Body's Master Plan," *New York Times*, 18 December 2001, from Sheryl Gay Stolberg, "Controversy Reignites Over Stem Cells and Clones," *New York Times*, 18 December 2001, and from Gina Kolata, "Company

Says It Produced Embryo Clones," *New York Times*, 26 November 2001.

3. Ernst & Young, *Focus on Fundamentals*, op. cit. note 1, pp. 9 and 17.

4. Ibid., p. 8.

5. Clyde Payne, The Catalyst Group, Spring House, PA, discussion with author, 13 December 2001.

6. Ibid.

7. Biotechnology Industry Organization (BIO), "Some Facts About Biotechnology," at <www.bio.org/er/statistics.asp>.

8. Ernst & Young, *European Life Sciences*, op. cit. note 1, p. 4.

9. Ibid., p. 8, and <finance.yahoo.com>.

10. Ernst & Young, *Life Sciences in France–2001* (New York: 2001), p. 8.

11. Ernst & Young, *European Life Sciences*, op. cit. note 1, p. 5; Carey Krause, "Germany Moves Closer to UK in Biotechnology," *Chemical Market Reporter*, 5 November 2001.

12. Jocelyn Kaiser, "Cuba's Billion-Dollar Biotech Gamble," *Science*, 27 November 1998, pp. 1626–28.

13. Crispin Thorold, "Indian Firms Embrace Biotechnology," *BBC News*, 6 April 2001; "India's Fermentation Queen," op. cit. note 1.

14. BIO, op. cit. note 7.

15. Ernst & Young, *Focus on Fundamentals*, op. cit. note 1.

16. Geoff Dyer and Victoria Griffith, "Growing Together," *Financial Times*, 19 December 2001.

17. Catalyst Group, op. cit. note 2.

18. Ibid.

19. BIO, op. cit. note 7.

20. U.N. Development Programme, *Human Development Report 1999* (New York: 2000), p. 67.

21. Ibid., p. 68.

22. ETC Group, *Globalization, Inc., Concentration in Corporate Power: The Unmentioned Agenda* (Winnipeg, MN, Canada: July/August 2001), p. 7.

23. "Human Genomes, Public and Private," *Nature*, 15 February 2001, p. 745.

24. Rosario Isasi, Health Law Department, Boston University, "Database of Global Policies on Human Cloning and Germ-line engineering," <www.glphr.org/genetic/genetic.htm>; Rick Weiss, "Scientists Declare Progress in Human Cloning," *Washington Post*, 8 August 2001.

25. Gerald Nadler, "U.N. Panel Backs Cloning Measure," *Associated Press*, 19 November 2001.

## APPLIANCE EFFICIENCY TAKES OFF
(pages 132–33)

1. Paul Waide, PW Consulting, e-mail to author, 11 January 2002. Sales growth reflects unit sales of cookers, clothes dryers, color televisions, dishwashers, freezers, fridge-freezers, refrigerators, and washing machines in 50 countries between 1994 and 1998. To prepare the global average, national market growth averages between 1994 and 1998 were weighted by cumulative sales in each country.
2. Peter du Pont, Danish Energy Management A/S, Bangkok, "Learning by Doing: The Wealth of Experience Implementing Standards and Labeling Programs in Asia," presented at Lessons Learned in Asia: Regional Conference on Energy Efficiency Standards and Labeling, organized by Collaborative Labeling and Appliance Standards Program (CLASP) and the U.N. Economic and Social Commission for Asia and the Pacific (ESCAP), Bangkok, 29–31 May 2001.
3. Ibid.
4. Ibid. Number of power plants is based on the assumption of 500-megawatt capacity for a medium-sized plant.
5. Steven Wiel and James E. McMahon, *Energy Efficient Labels and Standards: A Guidebook for Appliances, Equipment and Lighting* (Berkeley, CA: CLASP, Lawrence Berkeley National Laboratory, February 2001).
6. Ibid.
7. In 1992, the European Union had 12 members, but by the end of 1995, this had increased to 15. In early 2002, 13 countries have applications pending to join the EU. Information from European Union, at <www.en2001.se/static/eng/eu_info/utvidgning.asp>, viewed 31 January 2002.
8. Germany's Blue Angel, an eco-certification product label that is not specific to energy performance, was started in 1978; U.S. Environmental Protection Agency, *Environmental Labeling Issues, Policies, and Practices Worldwide* (Washington, DC: December 1998).
9. Energy Star certification is available for more than 30 different products; see <www.energystar.gov>.
10. U.S. Department of Energy (DOE), *Major Appliance Shopping Guide, 2002,* at <www.eren.doe.gov/consumerinfo/energy_savers/shoppingbody.html>.
11. N. Phumaraphand, "Evaluation Methods and Results of EGAT's Labeling Programs," presented at Lessons Learned in Asia: Regional Conference on Energy Efficiency Standards and Labeling, organized by CLASP and ESCAP, Bangkok, 29–31 May 2001.
12. Du Pont, op. cit. note 2.
13. Ibid.
14. Jennifer Thorne, American Council for an Energy-Efficient Economy (ACEEE), Washington, DC, discussion with author, 31 January 2002.
15. Ibid.
16. Worldwatch calculation based on DOE Web site concerning the U.S. refrigerator market at <www.eren.doe.gov/buildings/consumer_information/refrig/refwhy.html>, viewed 20 January 2002.
17. Steven Nadel, "Appliance and Equipment Efficiency Standards" (draft) (Washington, DC: ACEEE, 29 November 2001).
18. L. Greening et al., *Retrospective Analysis of National Energy-Efficient Standards for Refrigerators* (Berkeley, CA: Lawrence Berkeley National Laboratory, 1996); Wiel and McMahon, op. cit. 5.
19. Paolo Bertoldi, "European Union Effective Policies and Measures in Energy Efficiency in Appliances and Stand-by Losses Reduction," paper presented at the Chinese Energy-Efficient Appliances Conference, January 2002.
20. Ibid.
21. Wiel and McMahon, op. cit. note 5.
22. "What is CLASP?" February 2001, at <www.clasponline.org>.

## WATER STRESS DRIVING GRAIN TRADE
(pages 134–35)

1. Grain data from U.S. Department of Agriculture (USDA), *Production, Supply and Distribution,* electronic database, Washington, DC, updated December 2000.
2. Sandra Postel, "Water for Food Production: Will There Be Enough in 2025?" *BioScience,* August 1998, pp. 629–37.
3. Sandra Postel, *Pillar of Sand* (New York: W.W. Norton & Company, 1999).
4. Number water-stressed is a Global Water Policy Project (GWPP) calculation, based on population data from U.S. Bureau of the Census and runoff data from World Resources Institute et al., *World Resources 2000–01* (Washington, DC: World Resources Institute, 2000); import figure is from

USDA, op. cit. note 1, and represents annual net imports averaged over 1998–2000 in order to discount single-year anomalies.

5. USDA, op. cit. note 1.

6. GWPP, op. cit. note 4.

7. Ibid.

8. Sandra Postel, *Water Competition and Stress within Countries: Implications for Regional and Global Stability* (Amherst, MA: GWPP, May 2001).

9. Ibid.

10. Ibid.

11. Water deficit projection from Sandia National Laboratories, China Infrastructure Initiative: Decision Support Systems, at <www.igaia.sandia.gov/igaia/china/chinamodel.html>; grain analysis from Postel, op. cit. note 3, p. 76.

12. Postel, op. cit. note 3.

13. U.S. 2001 grain exports of 89 million tons from USDA, *Production, Supply, and Distribution*, electronic database, updated 11 December 2001.

14. L. Smedema, *Irrigation-Induced River Salinization: Five Major Irrigated Basins in the Arid Zone* (Colombo, Sri Lanka: International Water Management Institute, 2000); Postel, op. cit. note 3.

15. Gary Gardner, *Shrinking Fields: Cropland Loss in a World of Eight Billion*, Worldwatch Paper 131 (Washington, DC: Worldwatch Institute, July 1996).

16. Sandra Postel et al., "Drip Irrigation for Small Farmers: A New Initiative to Alleviate Hunger and Poverty," *Water International*, March 2001, pp. 3–13; see also Gary Gardner and Brian Halweil, *Underfed and Overfed: The Global Epidemic of Malnutrition*, Worldwatch Paper 150 (Washington, DC: Worldwatch Institute, March 2000).

## FOOD-BORNE ILLNESS WIDESPREAD
(pages 138–39)

1. World Health Organization (WHO), "Foodborne Diseases—Possibly 350 Times More Frequent Than Reported," press release (Geneva: 13 August 1997). Table 1 from the following: "*Camplyobacter Jejun*," in U.S. Food and Drug Administration (FDA) and Center for Food Safety and Applied Nutrition, *Bad Bug Book, Foodborne Pathogenic Microorganisms and Natural Toxins Handbook*, updated 8 March 2000, at <www.cfsan.fda.gov/~mow>, viewed 8 January 2002; listeria from U.K. Food Standards Agency, at <www.food.gov.

uk/healthiereating/foodrelatedconditions/foodpoisoning/foodbugs/Listeria>, viewed 3 January 2002; marine toxins, parasites, and *E. coli* from WHO, *Foodborne Disease: A Focus for Health Education* (Geneva: 2000); idem, "Multi-drug Resistant Salmonella typhimurium," Fact Sheet No. 139 (Geneva: January 1997).

2. WHO, "Food Safety and Foodborne Illness," Fact Sheet No. 237 (Geneva: September 2000).

3. Ibid.; WHO, *Foodborne Disease*, op. cit. note 1, p. 20.

4. WHO, "Food Safety—A Worldwide Public Health Issue," Fact Sheet, at <www.who.int/fsf/fctshtfs.htm>, viewed 31 December 2001.

5. WHO, *Foodborne Disease*, op. cit. note 1, p. 2.

6. Ibid.

7. Paul S. Mead et al., "Food-related Illness and Death in the United States," *Emerging Infectious Diseases*, September-October 1999, p. 607.

8. U.K. Food Standards Agency, "Food Poisoning," at <www.food.gov.uk/healthiereating/foodrelatedconditions/foodpoisoning/foodbornedisease>, viewed 11 January 2002; "Food Poisoning 'Rife and Under-reported,'" *BBC News*, 6 September 2000; WHO, *Foodborne Disease*, op. cit. note 1, p. 11; idem, "Foodborne Diseases," op. cit. note 1.

9. U.N. Food and Agriculture Organization (FAO), *The State of Food and Agriculture 2001* (Rome: 2001), p. 261.

10. Ewen C. D. Todd, "Foodborne and Waterborne Disease in Developing Countries—Africa and the Middle East," *Dairy and Environmental Sanitation*, February 2001, p. 110.

11. WHO, *Foodborne Disease*, op. cit. note 1.

12. Lawrence J. Dyckman, Director, Food and Agriculture Issues, Resources, Community, and Economic Development Division, U.S. Government Accounting Office, "Food Safety: Overview of Food Safety and Inspection Service and Food and Drug Administration Expenditures," 20 September 2000, p. 2.

13. U.K. Parliamentary Office of Science and Technology, *Safer Eating: Microbiological Food Poisoning and Its Prevention* (London: October 1997), p. 8.

14. Pan American Health Organization and Instituto PanAmericano de Proteccion de Alimentos y Zoonosis, "Inocuidad de Alimentos, Alimentos involucrados en los brotes," at <intranet.inppaz.org.ar/nhp/ehome.asp>, viewed 2 January 2002.

15. WHO, *Foodborne Disease*, op. cit. note 1, p. 8.

16. Geng-Sun Qian et al., "A Follow-up Study of Uri-

nary Markers of Aflatoxin Exposure and Liver Cancer Risk in Shanghai, People's Republic of China," *Cancer Epidemiology, Biomarkers, and Prevention*, January/February 1994; Dr. David Kensler, Johns Hopkins School of Public Health, discussion with author, 9 January 2002.

17. WHO, "Food Technologies and Public Health," at <www.who.int/fsf/Foodtech&publichealth.pdf>, viewed 15 January 2002

18. FAO, op. cit. note 9.

19. "Animal Factory Manure Discharge Tests at 1,900 Times State Maximum *E. Coli* Levels: Lenawee County Facility Already Under USEPA Order: Second Facility Nearby Has Massive Violation Following Day," press release (Washington, DC: Sierra Club, 27 December 2001); "Canadian Town Wary of Water," *Associated Press*, 20 December 2000.

20. F. Diez-Gonzalez et al., "Grain-feeding and the Dissemination of Acid-resistant *Escherichia coli* from Cattle," *US Dairy Forage Research Center, 1998 Research Summaries*, p. 64.

21. WHO, *Foodborne Disease*, op. cit. note 1, p. 26.

22. Nick Tattersall, "Stressed Farm Animals Contribute to Food Poisoning: U.K. study," *Manitoba Co-operator*, 15 March 2001.

23. Margaret Mellon, Charles Benbrook, and Karen Lutz Benbrook, *Hogging It: Estimates of Antimicrobial Use in Livestock* (Cambridge, MA: Union of Concerned Scientists, 2001), p. xi.

24. Michael Balter, "Uncertainties Plague Projections of vCJD Toll," *Science*, 26 October 2001, pp. 770–71.

25. "Ruling Signals Shift in French Mad Cow Cull Policy," *Reuters*, 10 January 2002.

26. "Tetrodotoxin," in FDA and Center for Food Safety and Applied Nutrition, op. cit. note 1.

27. "Monkey Brains Off the Menu in Central Africa," *Planet Ark*, 2 January 2001.

28. WHO, *Foodborne Disease*, op. cit. note 1, p. 21.

29. WHO Food Safety Programme, *Food Safety: An Essential Public Health Issue for the New Millennium* (Rome: 2000).

30. WHO, *Foodborne Disease*, op. cit. note 1, p. 23.

## SODA CONSUMPTION GROWS
(pages 140–41)

1. Beverage Marketing, *The Global Beverage Marketplace, 2001 Edition* (New York: 2001), p. 5. Data from 2000 are the most recent available globally; 1999 is the most recent year for national data. Note that carbonated soft drinks are a smaller category than "soft drinks," which include ready-to-drink teas and coffees, sports drinks, and fruitless "fruit drinks" and concentrates.

2. Beverage Marketing, op. cit. note 1, p. 5.

3. Ibid., p. 7.

4. U.S. Bureau of the Census, *International Data Base*, electronic database, updated 10 May 2000; Beverage Marketing, op. cit. note 1, p. 58.

5. John Rowdan, Beverage Marketing, e-mail to author, 23 January 2002.

6. Average U.S. soda consumption from Beverage Marketing, op. cit. note 1, p. 72; tap water from John Sicher, *Beverage Digest*, e-mail to author, 17 December 2001.

7. Census Bureau, op. cit. note 4; Beverage Marketing, op. cit. note 1, p. 53.

8. Beverage Marketing, op. cit. note 1, p. 67.

9. Ibid., p. 66.

10. Less than a quarter of carbonated soft drinks are diet, providing few to no calories; U.S. Department of Agriculture (USDA), Economic Research Service, *Per Capita Food Consumption Data System: Beverages*, at <www.ers.usda.gov/Data/Food ConsumptionSpreadsheets/beverage.xls>, viewed 13 January 2002, and Euromonitor International, "Soft Drinks: The International Market, 2001 Edition," from Trudy Griggs, e-mail to author, 19 December 2001. While diet soda does not provide calories, the safety of high-intensity sweeteners remains controversial, so consuming diet soda is not necessarily the healthier alternative.

11. USDA, op. cit. note 10.

12. Claude Cavadini et al., "U.S. Adolescent Food Intake Trends from 1965 to 1996," *Archives of Disease in Childhood*, vol. 83, no. 1 (2000), p. 19.

13. Carol Ballew et al., "Beverage Choices Affect Adequacy of Children's Nutrient Intakes," *Archives of Pediatrics and Adolescent Medicine*, November 2000, p. 1148.

14. National Osteoporosis Foundation, "Disease Statistics—Fast Facts," at <www.nof.org/osteoporosis/stats.htm>, viewed 23 January 2002.

15. Grace Wyshak, "Teenaged Girls, Carbonated Beverage Consumption, and Bone Fractures," *Archives of Pediatrics and Adolescent Medicine*, June 2000, p. 612.

16. David S. Ludwig et al., "Relationship Between

Consumption of Sugar-Sweetened Drinks and Childhood Obesity: A Prospective, Observational Analysis," *The Lancet*, 17 February 2001, p. 507. These drinks refer to ones with added sweeteners, not ones with naturally occurring sugars such as juices.

17. Ludwig et al., op. cit. note 16, p. 507.

18. U.S. Department of Health and Human Services, *The Surgeon General's Call to Action to Prevent and Decrease Overweight and Obesity, 2001*, at <www.surgeongeneral.gov/topics/obesity/calltoaction/CalltoAction.pdf>, viewed 16 January 2002.

19. Estimate derived from Beverage Marketing, op. cit. note 1, p. 58, from USDA, op. cit. note 10, and from using an average of 100 calories per 240 milliliter serving of non-diet soda; daily maximum from Judy Putnam et al., "Per Capita Food Supply Trends: Progress Toward Dietary Guidelines," *FoodReview*, September–December 2000, p. 12.

20. "Soft Drinks and Dental Caries: A Current Controversy," *The Colgate Oral Care Report*, <www.colgateprofessional.com/cp/ColPro.class/jsp/continuinged/OCR_11-3.pdf>, viewed 21 January 2002.

21. Euromonitor International, op. cit. note 10.

22. Roland R. Griffiths and Ellen M. Vernotica, "Is Caffeine a Flavoring Agent in Cola Soft Drinks?" *Archives of Family Medicine*, August 2000, p. 732.

23. National Soft Drink Association, at <www.nsda.org/WhatsIn/caffeinecontent.html>, viewed 21 January 2002; 34 milligrams is equivalent to about a third of a cup of coffee.

24. Ronald R. Watson, "Caffeine: Is it Dangerous to Health?" *American Journal of Health Promotion*, spring 1988; Roland Griffiths, Johns Hopkins University School of Medicine, e-mail to author, 22 January 2002.

25. Griffiths and Vernotica, op. cit. note 22, p. 731.

26. Ad Age Global, "Top Ten by Media Ad Spending Outside the US," at <www.adageglobal.com/cgi-bin/pages.pl?link=498>, viewed 21 January 2002.

27. Warner Brothers, press release, at <movies.warnerbros.com/pub/cmp/releases/cokehp.htm>, viewed 21 January 2002; SaveHarry.com, press release, at <www.saveharry.com/pressreleaseB.html>, viewed 21 January 2002.

28. Kate Zernike, "Coke to Dilute Push in Schools for Its Products," *New York Times*, 14 March 2001.

29. Coca-Cola North America, press release, at <www.twbg.com/whats_news/20010304Coke.html>, viewed 22 January 2002.

30. Coca-Cola Company, *2000 Annual Report*, at <annualreport2000.coca-cola.com/downloads/ko00ar.pdf> viewed 11 March 2002.

31. Barry M. Popkin and Colleen M. Doak, "The Obesity Epidemic Is a Worldwide Phenomenon," *Nutrition Reviews*, April 1998, pp. 106–14.

32. SaveHarry.com, op. cit. note 28.

33. Euromonitor International, "Marketing to Children: A World Survey, 2001 Edition," from Trudi Griggs, e-mail to author, 25 January 2002.

34. Ibid.

35. Michael F. Jacobson and Kelly D. Brownell, "Small Taxes on Soft Drinks and Snack Foods to Promote Health," *American Journal of Public Health*, June 2000, pp. 854–57.

## PREVALENCE OF ASTHMA RISING RAPIDLY (pages 142–43)

1. World Health Organization (WHO), *Bronchial Asthma*, Fact Sheet No. 206, January 2000, at <www.who.int/inf-fs/en/fact206.html>, viewed 30 October 2001.

2. Ibid.

3. Global Initiative for Asthma (GINA), *Global Strategy for Asthma Management and Prevention: NHLBI/WHO Workshop Report*, May 1996, p. vii; WHO, op. cit. note 1; GINA, *Global Strategy for Asthma Management and Prevention: National Institutes of Health*, rev. 2002, p. 12.

4. GINA (2002), op. cit. note 3, p. 12; WHO, *World Health Report 2001* (Geneva: 2001), pp. 144, 148.

5. GINA (2002), op. cit. note 3, p. 20–21.

6. Ibid.

7. National Institutes of Health, "NHLBI Reports New Asthma Data for World Asthma Day 2001: Asthma Still a Problem But More Groups Fighting It," press release (Bethesda, MD: 3 May 2001).

8. National Heart, Lung, and Blood Institute (NHLBI), "A Pocket Guide for Physicians and Nurses," Global Strategy for Asthma Management and Prevention, GINA, 1998, at <www.ginasthma.com/pocketguide/pocket.html>, viewed 14 January 2002.

9. GINA (1996), op. cit. note 3, p. 13.

10. U.S. Environmental Protection Agency (EPA), "Asthma & Upper Respiratory Illnesses," at <www.epa.gov/children/asthma.htm>, viewed 14 November 2001.

11. WHO, op. cit. note 1; immune system information from Dr. Patricia Noel, Division of Lung Diseases, NHLBI, discussion with author, 24 January 2002.

12. WHO, "Let Every Person Breathe: World Asthma Day," press release (Geneva: 2 May 2000).

13. GINA (1996), op. cit. note 3, p. 59.

14. WHO, op. cit. note 1; NHLBI, op. cit. note 6; GINA (1996), op. cit. note 3, pp. 35, 47; aspirin-specific information from Noel, op. cit. note 9.

15. The International Study of Asthma and Allergies in Childhood (ISAAC) Steering Committee, "Worldwide Variation in Prevalence of Symptoms of Asthma, Allergic Rhinoconjunctivitis, and Atopic Eczema: ISAAC," *Lancet*, 25 April 1998, p. 1225.

16. Ibid., p. 1231; Table 1 from Nadia Ait-Khaled, Donald Enarson, and Jean Bousquet, "Chronic Respiratory Diseases in Developing Countries: The Burden and Strategies for Prevention and Management," *Bulletin of the World Health Organization*, October 2001, p. 973.

17. ISAAC, op. cit. note 13, p. 1231.

18. U.S. Department of Health and Human Services, *Action Against Asthma: A Strategic Plan for the Department of Health and Human Services* (Washington, DC: May 2000); GINA, *Asthma Fact Sheet*, World Asthma Day 2002, at <207.159.65.33/wadsetup/materials/asthma.doc>, viewed 27 February 2002.

19. WHO, op. cit. note 1.

20. Ibid.

21. Ibid.

22. Ibid.

23. WHO, op. cit. note 10.

24. EPA, op. cit. note 8; WHO, op. cit. note 1.

25. C. Janson et al., "The European Community Respiratory Health Survey: What are the Main Results So Far?" *European Respiratory Survey*, vol. 18 (2001), p. 606.

26. "The Future of Asthma," *Lancet*, 18 October 1997, p. 1113.

27. WHO, op. cit. note 10; WHO, op. cit. note 1; GINA (1996), op. cit. note 3, p. 37.

28. WHO, op. cit. note 1; GINA (1996), op. cit. note 3, p. 43.

29. GINA (1996), op. cit. note 3, p. 31.

30. Erika von Mutius, "Towards Prevention," *Lancet*, 18 October 1997, Supplement 2, p. 1417; C. Janson et al., "Effect of Passive Smoking on Respiratory Symptoms, Bronchial Responsiveness, Lung Function, and Total Serum IgE in the European Community Respiratory Survey: a Cross-sectional Study," *Lancet*, 22/29 December 2001, pp. 2103, 2106, 2108.

31. WHO, op. cit. note 10.

32. "The Future of Asthma," op. cit. note 24; ISAAC, op. cit. note 13, p. 1231.

33. ISAAC, op. cit. note 13, pp. 1230–31; Corliss Karasov, "On a Different Scale: Putting China's Environmental Crisis in Perspective," *Environmental Health Perspectives*, October 2000, p. A-452; World Bank, *Health, Nutrition, and Population (HNP) Statistics*, at <devdata.worldbank.org/hnpstats/>, viewed 18 January 2002.

34. ISAAC, op. cit. note 13, p. 1231.

35. WHO, *Air Pollution*, Fact Sheet No. 187, September 2000, at <www.who.int/inf-fs/en/fact187.html>, viewed 22 January 2002.

36. ISAAC, op. cit. note 13, p. 1231; von Mutius, op. cit. note 28; GINA (1996), op. cit. note 3, p. 33.

37. GINA (1996), op. cit. note 3, p. 33; WHO, op. cit. note 1; UCB Institute of Allergy, European Allergy White Paper, 1999; von Mutius, op. cit. note 28.

38. William Booth, "Study: Pollution May Cause Asthma," *Washington Post*, 1 February 2002.

39. Anderson Wachira Kigotho, "Slums Bear the Brunt of African Asthma," *Lancet*, 20 September 1997, p. 874; GINA (1996), op. cit. note 3, p. 27; Margaret R. Becklake and Pierre Ernst, "Environmental Factors," *Lancet*, 18 October 1997, Supplement 2, p. 1013.

40. WHO, op. cit. note 10.

41. Ait-Khaled, Enarson, and Bousquet, op. cit. note 14, p. 977; WHO, *The WHO Model List of Essential Drugs*, at <www.who.int/medicines/organization/par/edl/infedlmain.htm>, viewed 25 January 2002.

42. Ait-Khaled, Enarson, and Bousquet, op. cit. note 14, pp. 975–77.

43. Janson et al., op. cit. note 28, p. 2108; WHO, op. cit. note 10.

44. Alan Macdermid, "Mites Bite Dust in Fight to Curb Asthma," (Glasgow) *The Herald*, 12 November 2001.

45. Booth, op. cit. note 36.

## MENTAL HEALTH OFTEN OVERLOOKED
(pages 144–45)

1. World Health Organization (WHO), *The World Health Report 2001* (Geneva: 2001), p. 19.

2. Ibid.
3. Ibid., p. 21; Table 1 from ibid., from National Institute of Mental Health, "Fact Sheets on Anxiety Disorders/Obsessive Compulsive Disorder," at <www.nimh.nih.gov/publicat/ocd.cfm> and <www.nimh.nih.gov/anxiety/anxiety/ocd/ocdfac.htm>, viewed 7 December 2001, from K. J. Neumaker, "Mortality and Sudden Death in Anorexia Nervosa," *International Journal of Eating Disorders*, April 1997, pp. 205–12, from P. F. Sullivan, "Mortality in Anorexia Nervosa," *American Journal of Psychiatry*, vol. 152, no. 7 (1995), pp. 1073–74, and from Merry N. Miller and Andres J. Pumariega, "Culture and Eating Disorders: A Historical and Cross-cultural Review," *Psychiatry*, summer 2001, pp. 98–103.
4. WHO, op. cit. note 1, p. 34.
5. Ibid., pp. 24–26.
6. Ibid.
7. Ibid.
8. Ibid.
9. U.S. Department of Health and Human Services, *Report of the Surgeon General's Conference on Children's Mental Health: A National Action Agenda* (Rockville, MD: 1999), p. 15.
10. WHO, op. cit. note 1, p. 36.
11. Ibid.
12. Ibid., p. 39; Ji Jianlin, "Committed Suicide in the Chinese Rural Areas," *Updates on Global Mental and Social Health*, Newsletter of the World Mental Health Project, Department of Social Medicine, Harvard Medical School, June 1999.
13. William Branigin and Leef Smith, "Mentally Ill Need Care, Find Prison," *Washington Post*, 25 November 2001; U.S. Department of Justice, Bureau of Justice Statistics, "Key Facts at a Glance, Correctional Populations 1980–2000," at <www.ojp.gov/bjs/glance/tables/corr2tab.htm>, viewed 7 December 2001.
14. Miller and Pumariega, op. cit. note 3, pp. 98–103.
15. H. H. Forster et al., "The Impact of Urbanization on Physical, Physiological and Mental Health of Africans in the North West Province of South Africa: The THUSA study," *South African Journal of Science*, September/October 2000, pp. 505–14.
16. WHO, op. cit. note 1, p. 30.
17. Ibid., p. 42.
18. Ibid.; J. H. Gold, "Gender Differences in Psychiatric Illness and Treatments: A Critical Review," *Journal of Nervous and Mental Diseases*, pp. 769–75.
19. WHO, *Mental Health Policy Project: Policy and Service Guidance Package* (Geneva: 2001), p. 10.
20. WHO, op. cit. note 1, p. 31.
21. Ibid., p. 32.
22. Ibid., p. x.
23. Ibid., p. 38–39.
24. Paul Gunderson et al., "The Epidemiology of Suicide Among Farm Residents or Workers in Five North-Central States, 1980–1988," *American Journal of Preventive Medicine*, May/June 1993, p. 26.
25. Ibid., p. 42.
26. "Domestic Violence Linked to Increased Suicide," *UN Wire*, 28 November 2001.
27. WHO, "Violence Against Women: A Priority Health Issue," *Violence Against Women Information Pack* (Geneva: July 1997).
28. IMS Health, *Pharmaceutical World Review* (London: 2001).
29. Shankar Vedantam, "Report Shows Big Rise in Treatment for Depression," *Washington Post*, 9 January 2002.
30. WHO, op. cit. note 1, p. 51.
31. Ibid., p. 16.

## POVERTY PERSISTS (pages 148–49)

1. The $1 a day is in 1993 purchasing power parity terms. The World Bank has used household survey data to estimate global income poverty since 1990, and has used historical surveys to derive estimates for 1987. The level just above "extreme poverty" is often called "overall poverty" or "relative poverty," and signifies insufficient income to meet essential non-food needs, such as for shelter and clothing.
2. World Bank, *World Development Report 2000/2001* (New York: Oxford University Press, 2000), p. 23.
3. Ibid.
4. International Forum for Agricultural Development, *Rural Poverty Report 2001: The Challenge of Ending Rural Poverty* (Oxford, U.K.: Oxford University Press, 2001); Dinesh Mehta, "Urbanization of Poverty," *Habitat Debate*, December 2000, pp. 1–3.
5. World Bank, op. cit. note 2, p. 25.
6. World Bank, op. cit. note 1.
7. Organisation for Economic Cooperation and Development, *Society at a Glance: OECD Social Indicators, 2001 Edition* (Paris: September 2001).

8. Barbara Ehrenreich, *Nickel and Dimed: On (Not) Getting By in America* (New York: Metropolitan Books, 2001); U.S. Census Bureau, "QT-03: Profile of Selected Economic Characteristics: 2000," *Census 2000 Supplementary Survey Summary Tables*, at <factfinder.census.gov/servlet/QTTable?ds_name =ACS_C2SS_EST_G00_&geo_id=01000US&qr_n ame=ACS_C2SS_EST_G00_QT02>, viewed 5 February 2002.

9. Amartya Sen, *Development as Freedom* (New York: Random House, 1999).

10. United Nations Development Programme (UNDP), *Human Development Report 2001* (New York: Oxford University Press, 2001).

11. Ibid.

12. Kirk Hamilton, "Focusing on Poverty and Environmental Links," in World Bank, *Environment Matters At the World Bank* (Washington, DC: 2001), p. 15.

13. Deepa Narayan et al., *Voices of the Poor: Crying Out for Change* (New York: Oxford University Press for the World Bank, 2000), p. 162.

14. Robert Hunter Wade, "The Rising Inequality of World Income Distribution," *Finance and Development*, December 2001, pp. 37–39.

15. World Bank, op. cit. note 1.

16. United Nations University, *World Income Inequality Database*, at <www.wider.unu.edu/wiid/wiid. htm>, viewed 6 January 2001.

17. Giovanni Andrea Cornia and Julius Court, *Inequality, Growth and Poverty in the Era of Liberalization and Globalization* (Helsinki, Finland: UNU World Institute for Development Economics Research, 2001).

18. Richard G. Wilkinson, *Unhealthy Societies: The Afflictions of Inequality* (New York: Routledge, 1996), pp. 72–109; Ichiro Kawachi, Bruce P. Kennedy, and Richard G. Wilkinson, *The Society and Population Health Reader: Income Inequality and Health* (New York: The New Press, 1999).

19. John W. Lynch et al., "Income Inequality and Mortality in Metropolitan Areas of the United States," in Kawachi, Kennedy, and Wilkinson, op. cit. note 18, pp. 69–81.

20. Thomas Homer-Dixon, *Environment, Scarcity and Violence* (Princeton, NJ: Princeton Press, 1999).

21. UNDP, *Overcoming Human Poverty: Poverty Report 2000* (New York: 2000), p. 19.

22. UNDP, op. cit. note 10.

23. Martin Ravallion, "Growth, Inequality and Poverty: Looking Beyond Averages," cited in Cornia and Court, op. cit. note 17.

24. Joshua Levin, "China's Divisive Development: Growing Urban-Rural Inequality Bodes Trouble," *Harvard International Review*, fall 2001, pp. 40–42; "Income Gap Continues To Widen Over Next Five Years," *China Population Today*, August 2001, p. 36.

25. U.S. Census data cited in Jared Bernstein and Lawrence Mishel, with research assistance by Thacher Tiffany, "Household Income Fails to Grow in 2000," *Economic Policy Institute Income Fax*, 25 September 2001; Nina Bernstein, "Widest Income Gap is Found in New York," *New York Times*, 19 January 2000; D'Vera Cohn and Sarah Cohen, "D.C. Gap Between Rich, Poor Widening: Census Data Show a City Polarized on Several Scales," *Washington Post*, 13 August 2001.

26. UNDP, *Human Development Report 1995* (New York: Oxford University Press, 1995).

27. UNDP, op. cit. note 10.

28. Robert Engelman, Brian Halweil, and Danielle Nierenberg, "Rethinking Population, Improving Lives," in Worldwatch Institute, *State of the World 2002* (New York: W.W. Norton & Company, 2002), pp. 127–48; Population Action International, "Fact Sheet: How Reproductive Health Services Work to Reduce Poverty" (Washington, DC: undated).

29. World Bank, *The East Asian Miracle* (London: Oxford University Press, 1993).

30. World Bank, *World Development Report 2002* (New York: Oxford University Press, 2002).

31. Cairo from James Drummond, "Freehold Provides Collateral for a Better Future," *Financial Times*, 16 October 2001; Danna Harman, "Kenya's Slums: New Political Battleground," *Christian Science Monitor*, 10 December 2001; Scott Baldauf, "Bombay's Poor Unmoved by Promise of Homes," *Christian Science Monitor*, 27 June 2001.

32. Hernando de Soto, *The Mystery of Capital: Why Capitalism Triumphs in the West and Fails Everywhere Else* (New York: Basic Books, 2000), pp. 18–20, 35.

## CAR-SHARING EMERGING (pages 150–51)

1. Worldwatch estimates of car sharers and number of vehicles based on e-mails from various car-sharing organizations worldwide to Erik Assadourian,

Worldwatch Institute, November and December 2001.

2. Katie Alvord, *Divorce Your Car: Ending the Love Affair with the Automobile* (Gabriola Island, BC, Canada: New Society Publishers, 2000), p. 174.

3. Peter Muheim & Partner, *CarSharing: The Key to Combined Mobility* (Lucerne, Switzerland: Das Aktionsprogramm Energie 2000, September 1998).

4. Alvord, op. cit. note 2.

5. Ibid.

6. Karl Steininger, Caroline Vogl, and Ralph Zettl, "Car-Sharing Organizations: The Size of the Market Segment and Revealed Change in Mobility Behavior," *Transport Policy*, vol. 3, no. 4 (1996), p. 178.

7. Germany from e-mails to Erik Assadourian, Worldwatch Institute, from Dirk Bake, Office Manager, Bundesverband CarSharing, 28 November 2001, and from Doris Johnsen, Marketing Coordinator, Stattauto, 21 December 2001; Switzerland from Cornelia Thoma, Support Assistant, Mobility CarSharing, e-mail to Erik Assadourian, Worldwatch Institute, 26 November 2001.

8. Growth in Washington and Boston from Mark Chase, transportation planner, Zipcar, e-mail to Erik Assadourian, Worldwatch Institute, 20 November 2001.

9. Steininger, Vogl, and Zettl, op. cit. note 6, pp. 177–78.

10. Peter Muheim & Partner, op. cit. note 3.

11. Ibid.

12. American Automobile Association, "Your Driving Costs," at <www.ouraaa.com/news/library/driving cost/driving.html>, viewed 16 December 2001.

13. Worldwatch estimate based on "Flexcar Pricing Schedule," at <www.flexcar.com/personal/pricing_de.asp>, viewed 8 March 2002.

14. Taylor Lightfoot Transport Consultants, *Pay As You Drive. Carsharing Final Report* (Ireland: EU-SAVE, 1997).

15. Peter Muheim & Partner, op. cit. note 3.

16. Ibid.

17. Worldwatch calculation based on information in ibid.

18. Peter Muheim & Partner, op. cit. note 3.

19. Mobility Car-Sharing, at <195.65.210.72/e/index.htm>, viewed 10 December 2001.

20. Peter Muheim & Partner, op. cit. note 3.

21. Susan A. Shaheen, "Carsharing in the United States: Examining Market Potential," Office of the Director, California Department of Transportation, Sacramento CA, undated, p. 2.

## SPRAWLING CITIES HAVE GLOBAL EFFECTS (pages 152–53)

1. Light rail from Terry Bronson, American Public Transit Association, e-mail to author, 8 March 2002; bicycles are Worldwatch estimate.

2. Jeffrey B. Kenworthy and Felix Laube et al., *An International Sourcebook of Automobile Dependence in Cities 1960–1990* (Boulder, CO: University Press of Colorado, 1999).

3. U.S. census data cited in David Rusk, *Inside Game, Outside Game: Winning Strategies for Saving Urban America* (Washington, DC: Brookings Institution Press, 1999), pp. 68–69.

4. U.S. census data cited in Leon Kolankiewicz and Roy Beck, *Weighing Sprawl Factors in Large U.S. Cities* (Arlington, VA: NumbersUSA, March 2001).

5. Goddard Space Flight Center, Scientific Visualization Studio, Greenbelt, MD, "Shenzhen, China Sprawl Effects," at <svs-f.gsfc.nasa.gov/imagewall/AAAS/china.html>, viewed 7 January 2002.

6. William Acevedo, Timothy Foresman, and Janis Buchanan, "Origins and Philosophy of Building a Temporal Database to Examine Human Transformation Processes," at <edcwww2.cr.usgs.gov/umap/pubs/asprs_wma.html>, viewed 1 May 1999.

7. Jeffrey Masek, University of Maryland, "Growth Patterns of Urban Sprawl," *Science Writers Guide to Landsat 7*, at <ltpwww.gsfc.nasa.gov/LANDSAT/>, viewed 21 April 1999.

8. Christopher Elvidge et al., "Satellite Inventory of Human Settlements Using Nocturnal Radiation Emissions: A Contribution for the Global Toolchest," *Global Change Biology*, October 1997, pp. 387–95.

9. Marc L. Imhoff et al., "Assessing the Impact of Urban Sprawl on Soil Resources in the United States Using Nighttime 'City Lights' Satellite Images and Digital Soils Maps," in Thomas D. Sisk, ed., *Perspectives on the Land Use History of North America: A Context for Understanding Our Changing Environment* (Washington, DC: U.S. Geological Survey, 1998), pp. 13–22.

10. U.S. estimate is for 1982–92 and is from American Farmland Trust, *Farming on the Edge* (Washington, DC: March 1997).

11. China estimate is for 1991–96 and is from Liu Yinglang, "Legislation to Protect Arable Land," *China Daily*, 15 September 1998.

12. J. T. Houghton et al., eds., *Climate Change 1995: The Science of Climate Change*, Contribution of Working Group I to the Second Assessment Report of the Intergovernmental Panel on Climate Change (Cambridge, U.K.: Cambridge University Press, 1996).

13. International Energy Agency, *$CO_2$ Emissions from Fuel Combustion, 1971–1998* (Paris: Organisation for Economic Co-operation and Development, 2000).

14. Jeffrey R. Kenworthy et al., *An International Sourcebook of Automobile Dependence in Cities, 1960–1990* (Boulder, CO: University Press of Colorado, 1999).

15. Consequences from R. T. Watson et al., eds., *Climate Change 1995: Impacts, Adaptations, and Mitigation*, Contribution of Working Group II to the Second Assessment Report of the Intergovernmental Panel on Climate Change (Cambridge, U.K.: Cambridge University Press, 1996); vulnerability of cities from Robert T. Watson et al., eds., *The Regional Impacts of Climate Change: An Assessment of Vulnerability*, A Special Report of IPCC Working Group II (Cambridge, U.K.: Cambridge University Press, 1998).

16. Watson et al., *Climate Change 1995: Impacts, Adaptations, and Mitigation*, op. cit. note 15.

17. Laurence Kalkstein and J. Scott Greene, "Evaluation of Climate/Mortality Relationships in Large U.S. Cities and the Possible Impacts of Climate Change," *Environmental Health Perspectives*, January 1997, pp. 84–93.

18. Rita Seethaler, *Health Costs Due to Road Traffic-related Air Pollution: An Impact Assessment Project of Austria, France and Switzerland*, Synthesis Report, prepared for the WHO Ministerial Conference on Environment and Health, London, June 1999, p. 9; Nino Künzli et al., "Public-Health Impact of Outdoor and Traffic-Related Air Pollution: A European Assessment," *Lancet*, 2 September 2000, pp. 795–801.

19. Traffic deaths from World Health Organization, *The World Health Report 1995* (Geneva: 1995), and from Christopher Willoughby, *Managing Motor-ization*, Discussion Paper, Transport Division (Washington, DC: World Bank, April 2000), p. ii.

20. Walter Hook and Michael Replogle, "Motorization and Non-Motorized Transport in Asia," *Land Use Policy*, vol. 13, no. 1 (1996), pp. 69–84.

21. Atlanta from Tim Lomax et al., *2001 Urban Mobility Study* (College Station, TX: Texas Transportation Institute, 2001); Bangkok from Richard Stren, "Transportation and Metropolitan Growth," in Richard Stren and Mila Freire, *The Challenge of Urban Government: Policies and Practices* (Washington, DC: World Bank Institute, 2001), p. 380.

22. United Nations, *World Urbanization Prospects: The 1999 Revision* (New York: 2000). Global urban population estimates are difficult to make, as the definition of "urban" and the reliability of census data vary from country to country. The U.N. figures cited here are for "urban agglomerations," which generally include the population in a city or town as well as that of adjacent suburbs.

23. Jos Dings, Centre for Energy Conservation and Environmental Technology, Delft, Netherlands, discussion with author, 22 February 2001; American Automobile Association, *Your Driving Costs, 1998* ed., cited in Stacy C. Davis, *Transportation Energy Data Book: Edition 19* (Oak Ridge, TN: Oak Ridge National Laboratory, September 1999), pp. 5-14–5-15.

24. United States from Mark M. Glickman, "Beyond Gas Taxes: Linking Driving Fees to Externalities," (San Francisco: Redefining Progress, March 2001), from Federal Highway Administration, *Addendum to the 1997 Federal Highway Cost Allocation Study, Final Report* (Washington, DC: U.S. Department of Transportation, Federal Highway Administration, May 2000), from Clifford Cobb, "The Roads Aren't Free," *Challenge*, May/June 1999, pp. 63–83, from U.S. Congress, Office of Technology Assessment, *Saving Energy in U.S. Transportation* (Washington, DC: 1994), from Todd Litman, "Transportation Cost Survey" (Victoria, BC, Canada: Victoria Transportation Policy Institute, 1992), from James MacKenzie, Roger Dower, and Donald Chen, *The Going Rate: What It Really Costs to Drive* (Washington, DC: World Resources Institute, 1992), and from James J. Murphy and Mark A. Delucchi, "A Review of the Literature on the Social Cost of Motor Vehicle Use in the United States," *Journal of Transportation and Statistics*, January 1998, pp. 14–42; Christopher

Zegras and Todd Litman, *An Analysis of the Full Costs and Impacts of Transportation in Santiago de Chile* (Washington, DC, and Santiago de Chile: International Institute for Energy Conservation, 1997) pp. 112–13.

25. VNG uitgeverij, *The Economic Significance of Cycling: A Study to Illustrate the Costs and Benefits of Cycling Policy* (The Hague: 2000), p. 42; Paul Guitink, Senior Transportation Specialist, World Bank, Washington, DC, discussion with author, 4 May 2001.

26. Taxes from Gerhard P. Metschies, *Fuel Prices and Taxation: With Comparative Tables for 160 Countries* (Eschborn, Germany: Deutsch Gesellschaft für Technische Zusammenarbeit, May 1999); trucks from Per Kågeson and Jos Dings, *Electronic Kilometre Charging for Heavy Goods Vehicles in Europe* (Brussels: European Federation for Transport and Environment, 1999); other pricing techniques from H. William Batt, "Motor Vehicle Transportation and Proper Pricing: User Fees, Environmental Fees, and Value Capture," *Ecological Economics Bulletin*, first quarter 1998, pp. 10–14; Ken Gwilliam and Zmarak Shalizi, "Road Funds, User Charges, and Taxes," *The World Bank Research Observer*, August 1999, pp. 159–85; David Weller, "For Whom the Road Tolls: Road Pricing in Singapore," *Harvard International Review*, summer 1998, pp. 10–11.

## TEACHER SHORTAGES HIT HARD
(pages 154–55)

1. UNESCO, "15 Million New Teachers Needed," World Teacher's Day, 5 October 2001, at <www.unesco.org/opi/eng/unescopress/2001/01-99e.shtml>, viewed 5 January 2002.

2. Regional breakdown from UNESCO, *World Education Indicators 2000*, Table 10, at <www.unesco.org/education/information/wer/htmlENG/tablesmenu.htm>, viewed 15 December 2001.

3. Vinod Thomas et al., *The Quality of Growth* (New York: Oxford University Press, 2000).

4. Harvard Law School, "Title I's Provisions Addressing the Quality of Teachers and Their Training Need to be Strengthened," at <www.law.harvard.edu/civilrights/conferences/SpecEd/Teachers.html>, viewed 5 January 2001.

5. National Council of Teachers in Mathematics, "Leave No Child Behind," at <www.nctm.org/news/president/2001-05president.htm>, viewed 11 January 2001.

6. The AIDS Campaign Team for Africa, *Exploring the Implications of the HIV/AIDS Epidemic for Educational Planning in Selected African Countries: The Demographic Question* (Washington, DC: World Bank, September 2000).

7. Ibid.

8. Desmond Cohen, *The HIV Epidemic and the Education Sector in Sub-Saharan Africa*, Issues Paper No. 32 (New York: U.N. Development Programme, 1999).

9. World Bank Education Draft for Comments, April 2001, p. 33, at <www.worldbank.org/education/globaleducationreform/pdf/Education%20Chapter%20Sourcebook.pdf>, viewed 4 January 2002.

10. "A Better World For All—Education 2000," at <www.paris21.org/betterworld/education.htm>, viewed 16 January 2002.

11. Ibid.

12. Elmo Frazer, "Exodus!! The Movement of Our Teachers," 18 June 2001, at <www.caribbeancity.com/news/commentary/news_commentary_18jun01.html>, viewed 15 December 2001.

13. Voluntary Services Overseas (VSO), "Education at Whose Expense? UK Teacher Shortages Filled at the Cost of the World's Poorest Children Warns VSO," at <www.vso.org.uk/media/aug2001_2.htm>, viewed 7 January 2002; "New York Needs 8000 Teachers by Fall," *CNN.com News*, 12 July 2001.

14. Oxfam International, *Education Charges: A Tax on Human Development* (Oxford: 12 November 2001).

15. Oxfam International, *Education Now: Briefing Notes on IMF Policies* (Oxford: 1999).

16. "Client Perspectives: Private Sector, Private Schools," *Impact*, summer 1999.

17. Julie Blair, "Softening Economy May Ease Hiring Crunch for Schools," *Education Week*, October 3, 2001, at <www.edweek.org/ew/newstory.cfm>, viewed 6 January 2002.

18. The Center for the Future of Teaching and Learning, "Teaching and California's Future: The Status of the Teaching Profession 2001," at <www.cftl.org/whatsnew.html>, viewed 8 February 2002.

19. "Dummkopf!" *The Economist*, 15 December 2001, p. 43.

20. National Education Association (NEA), "Teacher

Shortage Fact Sheet," at <www.nea.org/teaching/shortage.html>, viewed 15 December 2001.

21. Amanda Dunn, "Cuts in Training Numbers Blamed," (Melbourne) *The Age*, 22 November 2001.

22. "Dummkopf!" op. cit. note 19.

23. UNESCO, *World Education Report: Teachers and Teaching in a Changing World* (New York: United Nations, 1998).

24. UNESCO, Institute for Statistics (UIS), "Education For All 2000 Assessment Statistical Document," p. 44, at <www.unesco.org/statistics>, updated 4 January 2001.

25. UNESCO, op. cit. note 23.

26. U.S. Department of Education, at <www.ed.gov/databases/ERIC_Digests/ed436529.html>, viewed 5 January 2001.

27. UNICEF, *State of the World's Children 1999* (New York: 1999), p. 32.

28. Elaine Sciolino, "Radicalism: Is the Devil in the Demographics?" *New York Times*, 9 December 2001.

29. Save the Children UK, *Policy Paper: Education in Emergencies*, at <www.campaignforeducation.org>, viewed 1 January 2001.

30. Swedish International Development Cooperation Agency, "Development Co-operation with Cambodia," at <www.sida.se:80/Sida/jsp/Crosslink.jsp?d=369&a=9164>, viewed 5 January 2002.

31. U.S. Department of Education, op. cit. note 26.

32. UNESCO, *Gender Sensitive Education Statistics and Indicators: A Practical Guide, Training Materials for Workshops on Education Statistics and Indicators*, at <www.uis.unesco.org/en/pub/pub0.htm>, viewed 4 January 2001.

33. Ibid.

34. United Nations, "50th Anniversary of the Universal Declaration of Human Rights," at <www.un.org/rights/50/decla.htm>, viewed 30 January 2002.

## WOMEN SUBJECT TO VIOLENCE
(pages 156–57)

1. Lori Heise, Mary Ellsberg, and Megan Gottemoeller, "Ending Violence Against Women," *Population Reports*, December 1999, p. 1.

2. World Health Organization (WHO), *Violence Against Women*, WHO Consultation (Geneva: 1996).

3. Heise, Ellsberg, and Gottemoeller, op. cit. note 1, p. 5.

4. Table 1 from the following sources: female infanticide from Celia W. Dugger, "The Girls Who Don't Get Born," *New York Times*, 6 May 2001; female genital mutilation from WHO, "Estimated Prevalence Rates for FGM, Updated May 2001," at <www.who.int/frh-whd/FGM>, viewed 10 November 2001, and from Amnesty International, "Female Genital Mutilation: A Human Rights Information Pack," at <www.amnesty.org/ailib/intcam/femgen/fgm1.htm>, viewed 13 July 2001; rape from Patricia Tjaden and Nancy Thoennes, National Institute of Justice and Centers for Disease Control and Prevention, *Research In Brief*, November 1998, *Prevalence, Incidence, and Consequences of Violence Against Women: Findings from the National Violence Against Women Survey* (Washington, DC: U.S. Department of Justice, 1998), and from WHO Violence and Injury Prevention, "Violence Against Women: A Priority Health Issue," July 1997, at <www.who.int/violence_injury_prevention/vaw/infopack.htm>, viewed 13 November 2001; murders from UNICEF, "Domestic Violence Against Women and Girls," *Innocenti Digest*, May 2000; Molly Moore, "In Turkey, 'Honor Killing' Follows Families to Cities," *Washington Post*, 8 August 2001; Asian Human Rights Commission, "WHO Urges China to Reduce Female Suicides," *Human Rights Solidarity*, January 2000.

5. UNICEF, op. cit. note 4, p. 6.

6. O. P. Sharma, "2001 Census Results Mixed for India's Women and Girls," *Population Today*, May/June 2001.

7. John Pomfet, "In China's Countryside, 'It's a Boy,' Too Often," *New York Times*, 29 May 2001.

8. United Nations Statistics Division, "Indicators on Population," at <www.un.org/Depts/unsd/social/population.htm>, viewed 18 January 2002.

9. UNICEF, op. cit. note 5, p. 3.

10. Figure of 140 million from United Nations, "Genital Mutilation: WHO Official Blasts Practice at Ghana Meeting," *UN Wire*, 14 December 2001.

11. WHO, "Female Genital Mutilation," Fact Sheet No 241 (Geneva: June 2000).

12. Jennifer Loven, "Young Women More Often Victims of Domestic Violence, Study Finds," *Nando Times*, 29 October 2001.

13. Callie Marie Rennison, "Intimate Partner Violence and Age of Victim, 1993–1999," *Bureau of Justice*

*Statistics Special Report*, October 2001 (rev. 28 November 2001), p. 1.

14. Leonard J. Paulozzi et al., "Surveillance for Homicide Among Intimate Partners, United States, 1981–1998," *Mortality and Morbidity Weekly Review Surveillance Summaries*, 12 October 200.

15. Lori Heise, "Violence Against Women: Impact on Sexual and Reproductive Health," in Program for Appropriate Technology in Health, *Reproductive Health, Gender and Human Rights: A Dialogue* (Washington, DC: 2001), p. 42.

16. M. C. Ellsberg et al., "Candies in Hell: Women's Experience of Violence in Nicaragua," *Social Science & Medicine*, vol. 51 (2000), pp. 1601–02.

17. Noeleen Heyzer, "Violence Against Women: With an End in Sight," *Development Outreach* (World Bank Institute), spring 2001.

18. UNICEF, *State of the World's Children 2001* (New York: 2001).

19. Ibid.

20. Heyzer, op. cit. note 17.

21. UNICEF, op. cit. note 18.

22 "Mapping a Global Pandemic, Review of Current Literature on Rape, Sexual Assault and Sexual Harassment of Women Consultation on Sexual Violence Against Women," Global Forum for Health Research, 2000, at <www.globalforum health.org/Non_compliant_pages/vaw/litrevmain. html>, viewed 10 December 2001.

23. U.N. Population Fund, *The State of World Population 2000* (New York: 2000), p. 29.

24. Ibid.

25. Heise, Ellsberg, and Gottemoeller, op. cit. note 1, p. 25.

26. Ibid.

27. Suzan Fraser, "Suicides of Women Rising in Traditional Southeast Turkey," *Washington Post*, 9 November 2000.

28. Ibid.

29. Ibid.

30. Asian Human Rights Commission, op. cit. note 4.

31. United Nations, *Report of the Fourth World Conference on Women, Beijing 1995, Beijing Declaration and Platform for Action*, at <www.un.org/women watch/confer/beijing/reports/plateng.htm>, viewed 18 January 2002.

## VOTER PARTICIPATION DECLINES
(pages 158–59)

1. Rafael López Pintor and Maria Gratschew, eds., *Voter Turnout Since 1945: A Global Report* (Stockholm, Sweden: International Institute for Democracy and Electoral Assistance (IDEA), in press); Maria Gratschew, IDEA, e-mail to author, 13 February 2002.

2. Gratschew, op. cit. note 1.

3. Rafael López Pintor, Maria Gratschew, and Kate Sullivan, "Voter Turnout Rates from a Comparative Perspective," in López and Gratschew, op. cit. note 1, pp. 75–91.

4. Julie Ballington, "Youth Voter Turnout," in López and Gratschew, op. cit. note 1, pp. 111–14.

5. López, Gratschew, and Sullivan, op. cit. note 3.

6. IDEA, at <www.idea.int/voter_turnout>, viewed 4 February 2002.

7. Table 1 from López, Gratschew, and Sullivan, op. cit. note 3.

8. IDEA, "Voter Turnout," from Maria Gratschew, IDEA, e-mail to author, 16 January 2001.

9. López, Gratschew, and Sullivan, op. cit. note 3.

10. Ibid.

11. Freedom House, "World Population and Freedom Rating," at <www.freedomhouse.org/research/ freeworld/2001/population.htm>, viewed 4 February 2002.

12. Ibid.

13. Ibid.

14. Adrian Karatnycky, ed., *Freedom in the World: The Annual Survey of Political Rights and Civil Liberties, 2000–2001* (New York: Freedom House, 2001).

15. Ibid.

16. Rachel L. Swarns, "Zimbabwe's Political Turmoil Simmers Before Vote," *New York Times*, 5 March 2002; John Jeter, "Misery Grows as Zimbabwe Voting Nears," *Washington Post*, 3 March 2002.

17. Run-up to election from Danna Harman, "Kenya's Slums: New Political Battleground," *Christian Science Monitor*, 10 December 2001; historic context from Frank Holmquist and Ayuka Oendo, "Kenya: Democracy, Decline, and Despair," *Current History*, May 2001, pp. 201–06.

18. "Ghana's Peaceful Transfer of Power" (editorial), *The Guardian* (London), 5 January 2001; share of population from IDEA, op. cit. note 8.

19. Randolph Ryan, "Election Seen Ending Milosevic Era," *Boston Globe*, 23 December 2000; Blaine

Harden, "Milosevic is Accused, But All of Serbia Is On Trial," *New York Times*, 1 April 2001.

20. John Phillips and James Robson, "Democracy Faces Test in Troubled Kosovo," *Christian Science Monitor*, 16 November 2001; Cynthia Scharf, "Balkans; Belgrade May Steal Kosovo's Thunder," *Los Angeles Times*, 19 November 2000.

21. Luis Rubio, "The Americas: A Rule of Law Emerges in Mexico, Slowly," *Wall Street Journal*, 27 April 2001; Joseph L. Klesner, "The End of Mexico's One-Party Regime," *PS, Political Science and Politics*, March 2001, pp. 107–14.

22. David Gonzalez, "New Chance for Peru Chief to Take Reins," *New York Times*, 13 January 2002.

23. IDEA, op. cit. note 8.

24. Seth Mydans, "U.N. Certifies First Election In the Newly Born East Timor," *New York Times*, 11 September 2001.

25. "Election Results," *Elections Today* (newsletter of the International Foundation for Election Systems), spring 2001, p. 26.

26. Andrew Higgins, "Broken Ballot," *Wall Street Journal*, 21 December 2000.

27. Marjorie Miller, "Blair Sails to Second Election Landslide," *Los Angeles Times*, 8 June 2001.

28. Elizabeth Bomberg, *Green Parties and Politics in the European Union* (New York: Routledge, 1998).

29. Phyllis Myers, "Livability at the Ballot Box: State and Local Referenda on Parks, Conservation and Smarter Growth, Election Day 1998," Discussion Paper Prepared for the Brookings Institution Center on Urban and Metropolitan Policy, January 1999; idem, "Growth at the Ballot Box: Electing the Shape of Communities in November 2000," Discussion Paper Prepared for the Brookings Institution Center on Urban and Metropolitan Policy, November 2000.

## PROGRESS AGAINST LANDMINES
(pages 162–63)

1. Landmines here refers exclusively to anti-personnel mines, not anti-tank mines.

2. The treaty is formally known as the Convention on the Prohibition of the Use, Stockpiling, Production and Transfer of Anti-Personnel Mines and on Their Destruction; see United Nations, <domino.un.org/TreatyStatus.nsf>.

3. International Campaign to Ban Landmines (ICBL), *Landmine Monitor Report 2001: Executive Summary* (Washington, DC: August 2001), p. 5.

4. Don Hubert, *The Landmine Ban: A Case Study in Humanitarian Advocacy*, Occasional Paper No. 42 (Providence, RI: Thomas J. Watson Jr. Institute for International Studies, Brown University, 2000), pp. 7–27.

5. Ibid.

6. Human Rights Watch, *Landmine Use in Afghanistan*, Backgrounder (Washington, DC: October 2001), p. 4.

7. ICBL, op. cit. note 3, pp. 4–5.

8. Ibid., p. 11.

9. Ibid.

10. Ibid., pp. 12–13.

11. Ibid., p. 13.

12. Ibid., pp. 13–14.

13. Ibid., p. 14.

14. Ibid.

15. Ibid., p. 15.

16. Ibid., p. 14.

17. U.S. Department of State, *To Walk the Earth in Safety: The United States Commitment to Humanitarian Demining* (Washington, DC: November 2001), p. A-50.

18. ICBL, op. cit. note 3, p. 23.

19. Ibid.

20. Ibid., p. 26.

21. U.S. Department of State, op. cit. note 17, p. A-50.

22. ICBL, op. cit. note 3, p. 31.

23. Figure of 26,000 casualties estimate for earlier years from ibid.; 30,000 casualties estimate from U.N. General Assembly, "Assistance in Mine Clearance: Report of the Secretary-General," New York, 6 September 1994.

24. ICBL, op. cit. note 3, p. 31.

25. Ibid., p. 43.

26. Ibid.

27. *United Nations News*, at <www.un.org/News/dh/latest/page2.html#23>, 5 November 2001.

28. ICBL, op. cit. note 3, pp. 43–45.

29. U.S. Department of State, op. cit. note 17, Introduction and pp. iii–iv. The number of 40 countries includes support to Kosovo (legally part of the Federal Republic of Yugoslavia) and northwest Somalia (Somalia has lacked a central government for several years).

30. U.S. Department of State, op. cit. note 17, pp. iii–iv.

31. C. J. Chivers, "400 Demolition Experts Will Try to Harvest Afghanistan's Field of Mines," *New York*

# Notes

*Times*, 18 December 2001; U.S. Campaign to Ban Landmines, *Newsletter*, 5 November 2001.

32. Current State Department estimate from U.S. Department of State, op. cit. note 17, p. A-51; 1998 estimate reported in Human Rights Watch, op. cit. note 6, pp. 2–3.

33. Estimate of 1 million from Human Rights Watch, op. cit. note 6, pp. 2–3; Halo Trust, a British demining organization, estimates Afghanistan's landmines to number about 640,000, cited in Chivers, op. cit. note 31; higher estimate from U.S. Campaign to Ban Landmines, op. cit. note 31.

34. Human Rights Watch, op. cit. note 6, p. 2.

35. "Mines Plague Afghan Agriculture," *Arms Trade News*, August/September 2001, p. 3.

36. Human Rights Watch, op. cit. note 6, p. 3.

37. Ibid.

38. U.S. Department of State, op. cit. note 17, p. A-51.

39. Stephen Franklin, "10 Million Mines Makes Foot Travel a Risky Business," *Chicago Tribune*, 29 October 2001.

40. "Cambodia's Landmine Clearance Will Take 'Hundreds' of Years," *Arms Trade News*, August/September 2001, p. 3.

# THE VITAL SIGNS SERIES

Some topics are included each year in *Vital Signs*; others are covered only in certain years. The following is a list of topics covered in *Vital Signs* thus far, with the year or years they appeared indicated in parentheses. Those marked with a bullet (♦) appeared in Part One, which includes time series of data on each topic.

## AGRICULTURE and FOOD

*Agricultural Resources*
- ♦ Fertilizer Use (1992–2001)
- ♦ Grain Area (1992–93, 1996–97, 1999–2000)
- ♦ Grain Yield (1994–95, 1998)
- ♦ Irrigation (1992, 1994, 1996–99, 2002)
- Livestock (2001)
- Organic Agriculture (1996, 2000)
- Pesticide Control or Trade (1996, ♦2000, 2002)
- ♦ Pesticide Trade (2000)
- Transgenic Crops (1999–2000)
- Urban Agriculture (1997)

*Food Trends*
- ♦ Aquaculture (1994, 1996, 1998, 2002)
- Biotech Crops (2001–02)
- ♦ Cocoa Production (2002)
- ♦ Coffee (2001)
- ♦ Fish (1992–2000)
- ♦ Grain Production (1992–2002)
- ♦ Grain Stocks (1992–99)
- ♦ Grain Used for Feed (1993, 1995–96)
- ♦ Meat (1992–2000, 2002)
- ♦ Milk (2001)
- ♦ Soybeans (1992–2001)
- ♦ Sugar and Sweetener Use (2002)

## THE ECONOMY

*Resource Economics*
- ♦ Aluminum (2001)
- Arms and Grain Trade (1992)
- Commodity Prices (2001)
- Fossil Fuel Subsidies (1998)
- ♦ Gold (1994, 2000)
- Metals Exploration (1998, ♦2002)
- ♦ Metals Production (2002)
- ♦ Paper (1993, 1994, 1998–2000)
- Paper Recycling (1994, 1998, 2000)
- ♦ Roundwood (1994, 1997, 1999, 2002)
- Seafood Prices (1993)
- ♦ Steel (1993, 1996)
- Steel Recycling (1992, 1995)
- Subsidies for Environmental Harm (1997)
- Wheat/Oil Exchange Rate (1992–93, 2001)

*World Economy and Finance*
- ♦ Agricultural Trade (2001)
- Aid for Sustainable Development (1997, 2002)
- ♦ Developing-Country Debt (1992–95, 1999–2002)
- Environmental Taxes (1996, 1998, 2000)
- Food Aid (1997)
- ♦ Global Economy (1992–2002)
- Microcredit (2001)
- ♦ Oil Spills (2002)

Private Finance in Third World (1996, 1998)
R&D Expenditures (1997)
Socially Responsible Investing (2001)
Stock Markets (2001)
♦ Trade (1993–96, 1998–2000, 2002)
Transnational Corporations (1999–2000)
♦ U.N. Finances (1998–99, 2001)

*Other Economic Topics*
♦ Advertising (1993, 1999)
Charitable Donations (2002)
Cigarette Taxes (1993, 1995, 1998)
Ecolabeling (2002)
Government Corruption (1999)
Health Care Spending (2001)
Pharmaceutical Industry (2001)
PVC Plastic (2001)
Satellite Monitoring (2000)
♦ Storm Damages (1996–2001)
♦ Television (1995)

## ENERGY and ATMOSPHERE

*Atmosphere*
♦ Carbon Emissions (1992, 1994–2002)
♦ CFC Production (1992–96, 1998, 2002)
♦ Global Temperature (1992–2002)

*Fossil Fuels*
♦ Carbon Use (1993)
♦ Coal (1993–96, 1998)
♦ Fossil Fuels Combined (1997, 1999–2002)
♦ Natural Gas (1992, 1994–96, 1998)
♦ Oil (1992–96, 1998)

*Renewables, Efficiency, Other Sources*
♦ Compact Fluorescent Lamps (1993–96, 1998–2000, 2002)
♦ Efficiency (1992)
♦ Geothermal Power (1993, 1997)
♦ Hydroelectric Power (1993, 1998)
♦ Nuclear Power (1992–2002)
♦ Solar Cells (1992–2002)
♦ Wind Power (1992–2002)

## THE ENVIRONMENT

*Animals*
Amphibians (1995, 2000)
Aquatic Species (1996, 2002)
Birds (1992, 1994, 2001)
Marine Mammals (1993)
Primates (1997)
Vertebrates (1998)

*Natural Resource Status*
Coral Reefs (1994, 2001)
Farmland Quality (2002)
Forests (1992, 1994–98, 2002)
Groundwater Quality (2000)
Ice Melting (2000)
Ozone Layer (1997)
Water Scarcity (1993, 2001–02)
Water Tables (1995, 2000)
Wetlands (2001)

*Natural Resource Uses*
Biomass Energy (1999)
Dams (1995)
Ecosystem Conversion (1997)
Energy Productivity (1994)
Organic Waste Reuse (1998)
Soil Erosion (1992, 1995)
Tree Plantations (1998)

*Pollution*
Acid Rain (1998)
Algal Blooms (1999)
Forest Damage from Air Pollution (1993)
Lead in Gasoline (1995)
Nuclear Waste (1992, ♦1995)
Pesticide Resistance (♦1994, 1999)
♦ Sulfur and Nitrogen Emissions (1994–97)
Urban Air Pollution (1999)

*Other Environmental Topics*
Environmental Treaties (♦1995, 1996, 2000, 2002)
Nitrogen Fixation (1998)
Pollution Control Markets (1998)
Semiconductor Impacts (2002)
Transboundary Parks (2002)

## THE MILITARY

- ♦ Armed Forces (1997)
  Arms Production (1997)
- ♦ Arms Trade (1994)
  Landmines (1996, 2002)
- ♦ Military Expenditures (1992, 1998)
- ♦ Nuclear Arsenal (1992–96, 1999, 2001)
  Peacekeeping Expenditures (1993, ♦1994–2002)
- ♦ Wars (1995, 1998–2002)
  Small Arms (1998–99)

## SOCIETY and HUMAN WELL-BEING

*Health*
- ♦ AIDS/HIV Incidence (1994–2002)
  Asthma (2002)
  Breast and Prostate Cancer (1995)
- ♦ Child Mortality (1993)
- ♦ Cigarettes (1992–2001)
  Drug Resistance (2001)
  Endocrine Disrupters (2000)
  Hunger (1995)
- ♦ Immunizations (1994)
- ♦ Infant Mortality (1992)
  Infectious Diseases (1996)
  Life Expectancy (1994, ♦1999)
  Malaria (2001)
  Malnutrition (1999)
  Noncommunicable Diseases (1997)
  Obesity (2001)
- ♦ Polio (1999)
  Safe Water Access (1995)
  Sanitation (1998)
  Soda Consumption (2002)
  Traffic Accidents (1994)
  Tuberculosis (2000)

*Reproduction and Women's Status*
  Family Planning Access (1992)
  Female Education (1998)
  Fertility Rates (1993)
  Maternal Mortality (1992, 1997)
- ♦ Population Growth (1992–2002)
  Sperm Count (1999)

Violence Against Women (1996, 2002)
Women in Politics (1995, 2000)

*Social Inequities*
  Homelessness (1995)
  Income Distribution (1992, 1995, 1997, 2002)
  Language Extinction (1997, 2001)
  Literacy (1993, 2001)
  Prison Populations (2000)
  Social Security (2001)
  Teacher Supply (2002)
  Unemployment (1999)

*Other Social Topics*
  Aging Populations (1997)
  Fast-Food Use (1999)
  Nongovernmental Organizations (1999)
  Refugees (♦1993–2000, 2001)
  Religious Environmentalism (2001)
  Urbanization (♦1995–96, ♦1998, ♦2000, 2002)
  Voter Turnouts (1996, 2002)
  Wind Energy Jobs (2000)

## TRANSPORTATION and COMMUNICATION

- ♦ Air Travel (1993, 1999)
- ♦ Automobiles (1992–2002)
- ♦ Bicycles (1992–2002)
  Car-Sharing (2002)
  Computer Production and Use (1995)
  Gas Prices (2001)
  Electric Cars (1997)
- ♦ Internet (1998–2000, 2002)
- ♦ Motorbikes (1998)
- ♦ Railroads (2002)
- ♦ Satellites (1998–99)
- ♦ Telephones (1998–2000, 2002)
- ♦ Tourism (2000)
  Urban Transportation (1999, 2001)

# WORLDWATCH PUBLICATIONS

**WORLD WATCH** This award-winning, bimonthly magazine keeps you up-to-speed on the latest developments in global environmental trends. *One year (6 issues) $25 individual, $27 institution, $15 student (outside North America: $40 individual, $42 institution, $30 student).*

**State of the World** Worldwatch's flagship annual is the most widely used public policy analysis in any field. The 2002 Special World Summit Edition focuses on the agenda of the upcoming World Summit on Sustainable Development, which will assess progress on global environmental issues. *($15.95 plus shipping and handling.)*

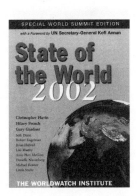

**State of the World Library** With this unique subscription package, you will receive *State of the World* <u>and</u> all five *Worldwatch Papers* as they are published throughout the year. *One year subscription: $39 individual, $43 institution, $30 student (outside North America: $49 individual, $53 institution, $45 student).*

**Be Sure to Visit Our Web Site (www.worldwatch.org)** Visit www.worldwatch.org for more information on the Worldwatch Institute, or to order any of the above publications. You may also contact us by mail, phone, fax, or e-mail.

## 4 Easy Ways to Order

1. Mail: Worldwatch Institute, P.O. Box 188, Williamsport, PA  17703-9913 USA
2. Call: (888) 544-2303 or (570) 320-2076
3. Fax: (570) 320-2079
4. E-mail: wwpub@worldwatch.org

*The Worldwatch Institute is a nonprofit 501(c)(3) public interest research organization and welcomes your tax-deductible contribution to advance its work.*

**WORLDWATCH INSTITUTE**
1776 Massachusetts Ave., NW
Washington, DC  20036
www.worldwatch.org

R Coddington    Food          Forestry
                Media         media
                9/10          15/15


A. MacMillan    News 7.5